T0296232

Spatiotemporal Analysis of Extreme Hydrological Events

Spatiotemporal Analysis of Extreme Hydrological Events

Edited by

Dr Ir. Gerald Corzo

Senior Researcher
Chair Group of Hydroinformatics
Department of Integrated Water Systems and Governance
IHE Delft Institute of Water Education
Delft, The Netherlands

Emmanouil A. Varouchakis

School of Environmental Engineering
Technical University of Crete
Chania, Greece

Elsevier
Radarweg 29, PO Box 211, 1000 AE Amsterdam, Netherlands
The Boulevard, Langford Lane, Kidlington, Oxford OX5 1GB, United Kingdom
50 Hampshire Street, 5th Floor, Cambridge, MA 02139, United States

Library of Congress Cataloging-in-Publication Data
A catalog record for this book is available from the Library of Congress

British Library Cataloguing-in-Publication Data
A catalogue record for this book is available from the British Library

ISBN: 978-0-12-811689-0

For information on all Elsevier publications visit our website at
https://www.elsevier.com/books-and-journals

Publisher: Candice Janco
Acquisition Editor: Louisa Hutchins
Editorial Project Manager: Emily Thomson
Production Project Manager: Denny Mansingh
Cover Designer: Matthew Limbert

Typeset by TNQ Technologies

"To my wife K. S. and son A. E. V"

Contents

List of Contributors

Gabriela Alvarez-Mieles IHE Delft Institute of Water Education, Delft, The Netherlands; Delft University of Technology, Faculty CiTG, Delft, The Netherlands; Universidad de Guayaquil, Facultad de Ciencias Naturales, Guayaquil, Ecuador

Biswa Bhattacharya IHE Delft Institute of Water Education, Delft, The Netherlands

Gerald Corzo UNESCO-IHE Delft Institute of Water Education, Delft, The Netherlands

Vitali Diaz UNESCO-IHE Institute for Water Education, Delft, the Netherlands; Water Resources Section, Delft University of Technology, Delft, the Netherlands

Miguel Laverde-Barajas IHE Delft Institute of Water Education, Delft, The Netherlands; Delft University of Technology, Water Resources Section, Delft, The Netherlands

Hung Manh Le National Central for Water Resources Planning and Investigation (NAWAPI), Ministry of Natural Resources and Environment (MONRE) of Vietnam, Hanoi, Vietnam

Vicente Medina Thermal Engines Department, Technical University of Catalonia, Barcelona, Spain

Arthur E. Mynett IHE Delft Institute of Water Education, Delft, The Netherlands; Delft University of Technology, Faculty CiTG, Delft, The Netherlands

Bang Luong Nguyen Faculty of Water Resources Engineering, Thuy Loi University, Hanoi, Vietnam

José R. Pérez Instituto Nacional de Recursos Hidráulicos (INDRHI), Santo Domingo, Dominican Republic

Dimitri P. Solomatine UNESCO-IHE Institute for Water Education, Delft, the Netherlands; Water Resources Section, Delft University of Technology, Delft, the Netherlands

Remko Uijlenhoet Hydrology and Quantitative Water Management Group, Wageningen University, Wageningen, The Netherlands

Henny A.J. Van Lanen Hydrology and Quantitative Water Management Group, Wageningen University, Wageningen, The Netherlands

Emmanouil A. Varouchakis School of Environmental Engineering, Technical University of Crete, Chania, Greece

Preface

Environmental and hydrological problems are spatial and temporal in nature. Spatiotemporal processes in physically based models provide a significant amount of information that is seldom analyzed or understood. Past experiences of science such as hydrology relied on a few measurements and empirical assumptions of water resources processes. Satellite information, spatially distributed models, and widely available global information have opened an essential area of development. For this, new methodologies have been studied, from the concepts of space/time statistics to deterministic machine learning and data-driven methods. Spatiotemporal analysis identifies and explains large-scale anomalies, which are useful for understanding hydrological characteristics and subsequently predicting hydrological events, even redefining the concepts of an event such that it can be characterized in multiple dimensions (space, time, and others). Methodologies that explore these spatial and temporal dimensions are critical in the genesis of this book.

Aside from this, hydrological problems have essential characteristics in their spatial and temporal dimensions at the same time. Analysis of hydrological extremes is essential, and most studies have reached the level of generalizing the problems to be able to adapt results to the available data. When new data arrive, a process to narrow the key variables is required to solve or understand the problem. The study of hydrological extremes requires advanced spatiotemporal methods to be able to narrow those studies and to extend analysis and prediction, including estimating the probability of their occurrence and the associated risk. Critical hydrological issues are extreme hydrological phenomena (e.g., precipitation, runoff), floods, low flows, and droughts. However, concepts are not just limited to this type of extreme, but also to its spatiotemporal consequences, such as the changes in a habitat of species due to flooding. The spatiotemporal study of extreme hydrological events at the watershed scale aids in understanding the relationship between their magnitude and the probability of these events occurring.

This book starts with the development of geostatistics. This discipline investigates the statistics of spatially and temporally extended variables. Spatiotemporal analysis and geostatistics are at the forefront of research these days, and their impact is expected to increase in the future. In this book, we follow the trend that is driven by increasing needs to advance risk assessment and management of strategies for extreme events. Current trends and variability of hydrological topics call for novel approaches of spatiotemporal and geostatistical analysis to assess, predict, and manage water-related topics.

Spatial statistics/geostatistics is used to map spatial observations from monitoring stations for hydrological or water resources data, assess spatial data quality, relating the accuracy of spatial data to their intended use, sample design optimization, model dependency structures, and draw valid inferences from a limited set of spatial data in agriculture, hydrology, hydrogeology, soil science, and ecology. Geostatistical models have been widely used in water resources management projects to represent and predict the spatial variability of aquifer levels. Also, they can be applied as surrogates to numerical models when the available hydrogeological data are scarce. For space/time data, spatiotemporal geostatistical approaches can model the examined variability by incorporating the compound space/time correlations.

The aim of this book is to provide a valuable reference of well-defined and innovative methodologies of spatiotemporal analysis in a hydrological context. The book is not limited to geostatistics but extends its concepts to pattern recognition algorithms used to map spatial dimensions into temporal variables. Multiple techniques are explained and described using real-life examples from different countries in the world.

Dr. Gerald A. Corzo Perez
Dr. Emmanouil A. Varouchakis

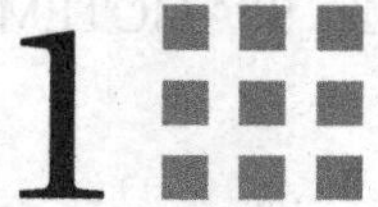

Geostatistics: Mathematical and Statistical Basis

Emmanouil A. Varouchakis
SCHOOL OF ENVIRONMENTAL ENGINEERING, TECHNICAL UNIVERSITY OF CRETE, CHANIA, GREECE

1. Random Fields

Geostatistics is intrinsically connected and based on the mathematical concept of random fields (RFs). RFs can be considered a set of random variables that describe the spatiotemporal variation of a physical variable size (e.g., hydraulic head, concentration of a pollutant). Contrary to functions that have a specific mathematical expression, RFs do not have a specific expression that represents all possible states. Each state is one sample of the field and is characterized by a probability determined by the multidimensional probability density function (PDF) of the field. Therefore an RF can be considered as a multidimensional random variable. Due to the interdependence of the physical characteristics in different points in space, RFs have particular mathematical properties that distinguish them from a set of independent random variables.

There are various categories of RFs. If the field takes values only from a finite set of numbers, it is called a discrete field. If the values of the field belong to a continuous interval of real numbers, the field is called a continuous field. When variation is defined in a continuous space, such as natural fields, a continuous field is created. On the contrary, when the positions of a grid are defined the field is called a lattice field.

Lattice fields are used in computational (e.g., simulation of the distribution of contaminants in groundwater) and theoretical studies, because grid symmetry allows the use of efficient numerical methods (e.g., fast Fourier transform). Moreover, lattice fields allow benchmarking of different geostatistical methods.

In practice the measurements represent a finite number of points, the distribution of which does not necessarily have the symmetry of a regular grid. In these cases the network of sampling points is inhomogeneous. The terms disordered lattice and off-lattice can be used as well. In such cases, geostatistical methods are needed to operate adequately, considering the limitations of each spatial distribution. If the distribution is off-lattice, the evaluation or simulation procedure is realized on a gridded background that covers the area of interest.

Spatiotemporal Analysis of Extreme Hydrological Events. https://doi.org/10.1016/B978-0-12-811689-0.00001-X

The concept of RFs is based on two key terms: randomness and interdependence of values of physical quantities at different points of the space. Randomness characterizes phenomena in which knowledge of a situation with complete accuracy is impossible due to various constraints. Such constraints originate from the variability of different physical quantities in space and the uncertainty due to the limited number of measurements. In these cases the result (the value of the phenomenon) is determined via a probability distribution function, which defines the probability of occurrence of each state.

Spatial dependence is a particular feature in random fields and describes the reliance between the values of two different points in the field. The probability distribution of the field embodies correlations between different points, so the probability of observing a value at a point depends on the values in adjacent points (Christakos and Hristopulos, 1998; Chiles and Delfiner, 1999).

2. Basic Concepts in Random Fields

An RF is denoted as $Z(\mathbf{s})$, where $\mathbf{s}$ is a position vector $\mathbf{s} = (x,y)$. $Z(\mathbf{s})$ represents all possible states in the field, while $z(\mathbf{s})$ denotes the values that correspond to a specific state. PDF of the field is denoted as $f_Z[z(\mathbf{s})]$. Index Z indicates the field, while the argument of the function is the values of the state of the field (e.g., hydraulic head, concentration of pollutants).

The PDF of a random field includes all values in the space where the field is defined. Therefore PDF is common for any number of points. One-dimensional or point PDF describes all possible states in the field on a specific point. It is possible that the one-dimensional PDF changes from point to point and that happens when the field is inhomogeneous. Proportionally, a two-dimensional PDF of the field expresses the interdependence of possible states of two points, while a multidimensional PDF describes the interdependence of all possible situations for N points.

Another type of function that provides information about the properties of a random field is statistical moments. Statistical moments are deterministic functions that represent average values in all possible situations. In practice, usually a low order (up to second order) statistical moment, as mean value, dispersion, and covariance functions and a semivariogram are useful (Goovaerts, 1997).

Spatial random fields (SRFs) are random fields whose location plays the primary role when the property values are spatially correlated. An SRF state can be decomposed into a deterministic trend $m_Z(\mathbf{s})$, a correlated fluctuation $Z'_\lambda(\mathbf{s})$, and an independent random noise term $e(\mathbf{s})$, so that $Z(\mathbf{s}) = Z'_\lambda(\mathbf{s}) + m_Z(\mathbf{s}) + e(\mathbf{s})$. The fluctuation term corresponds to "fast variations" that reveal structure at small scales, which nonetheless exceeds a cut-off λ; the trend is often determined from a single available realization. Random noise represents nonresolved inherent variability due to resolution limits, purely random additive noise, or nonsystematic measurement errors. The classical approach of SRFs is based on Gaussian SRFs (GSRFs) and various generalizations for non-Gaussian distributions (Wackernagel, 2003). The covariance matrix therefore is used to determine the spatial structure for the GSRFs, which is estimated from the distribution of the data in space. Generally, SRF model spatial correlations of variables have various applications, e.g., in hydrology (Kitanidis, 1997), environmental pollutant mapping, risk assessment (Christakos, 1991), mining exploration, and reserves estimation (Goovaerts, 1997).

2.1 Mean Value

The mean value of a random field is given by:

$$m'_Z(\mathbf{s}) = E[Z(\mathbf{s})] \tag{1.1}$$

where $E[Z(\mathbf{s})]$ denotes the mean value calculated in all states of the field, i.e.:

$$E[Z(\mathbf{s})] = \int dz\, f_Z(z; \mathbf{s})z \tag{1.2}$$

where z are the values that correspond to a given state. The integral limits depend on the space where field Z is defined. If the field takes all negative and positive values the integral varies from $-\infty$ to ∞. If the field takes only positive values the integral ranges from 0 to ∞. If it is known that the values of the field are limited to a predetermined interval [a,b], the integral is calculated in this interval. In the latter equation it can be noted that the average value may depend on position **s**, which comes from a possible dependence between the one-dimensional PDF and the position. Since PDF is not always known in advance, mean value is estimated through the sample using statistical methods. This is the average of all values in the sample (Hristopulos, 2008): $\widehat{m'_Z(\mathbf{s})} = \frac{1}{N}\sum_{i=1}^{N} z_i(\mathbf{s})$. A useful application topic of the mean value is to describe the large-scale trends in a random field. Mean value $m'_Z(\mathbf{s})$ is defined using reference functions. They can be divided into general and local dependence patterns. In the case of general dependence, only one mathematical equation describes the variance in the entire area. This kind of dependence pattern could be a linear dependence, which expresses the existence of a constant slope, a polynomial dependence, a periodic dependence, or the overlay of two or more patterns, e.g., a polynomial and a periodic. In cases where the general dependence patterns are insufficient for the exact determination of the trends, the use of local dependence functions is preferable (e.g., local polynomials). Such a type of dependence is used in the model of locally weighed regression (Isaaks and Srivastava, 1989).

2.2 Variance

Variance in a random field is given by the mean value of the squared fluctuation according to the equation:

$$\sigma_Z^2(\mathbf{s}) \equiv E\left[\{Z(\mathbf{s}) - m'_Z(\mathbf{s})\}^2\right] = E\left[\widetilde{Z}^2(\mathbf{s})\right]. \tag{1.3}$$

In general, it is possible for the variance to vary from point to point while remaining stable only when the field is statistically homogeneous. The variance fluctuations in an RF mean that the fluctuations in the field change from point to point (Isaaks and Srivastava, 1989).

2.3 Covariance Function

Another property that gives useful information for an RF is the centered covariance function (CCF), which is defined as (Isaaks and Srivastava, 1989):

$$c_Z(\mathbf{s}_1, \mathbf{s}_2) \equiv E[\{Z(\mathbf{s}_1) - m'_Z(\mathbf{s}_1)\}\{Z(\mathbf{s}_2) - m'_Z(\mathbf{s}_2)\}]. \tag{1.4}$$

The RF $\widetilde{Z}(\mathbf{s}_1) \equiv Z(\mathbf{s}_1) - m'_Z(\mathbf{s}_1)$ corresponds to the fluctuation in field $Z(\mathbf{s}_1)$ around the mean value at point $\mathbf{s}_1$. The mean value of the fluctuation field is equal to zero, $E[\widetilde{Z}(\mathbf{s}_1)] = 0$. Based on the previous equations it holds:

$$c_Z(\mathbf{s}_1, \mathbf{s}_2) = E[\widetilde{Z}(\mathbf{s}_1)\widetilde{Z}(\mathbf{s}_2)]. \tag{1.5}$$

Specifically, CCF describes quantitatively the dependence of the fluctuations between two different points in the field. When the points of the covariance function coincide, the value is equal to the variance of the field at that point, $c_Z(\mathbf{s}_1, \mathbf{s}_1) = \sigma^2_Z(\mathbf{s}_1)$. On the contrary, when the distance between two points increases, the dependence on the fluctuations is reduced (Cressie, 1993).

In geostatistical analysis the experimentally determined spatial dependence is fitted to an optimal model selected by a set of accepted theoretical functions (e.g., exponential, Gaussian, power law, etc.). A function is a valid covariance function if and only if it satisfies the following criteria:

$$\sum_{i=1}^{N}\sum_{j=1}^{N} a_i a_j c_Z(\mathbf{s}_i - \mathbf{s}_j) \geq 0, \tag{1.6}$$

for any real weights a, $i,j = 1,\ldots,N$ and any positive integer N. Acceptance conditions are also necessary for the covariance function. The acceptance conditions are set by Bochner's theorem (Bochner, 1959). This is expressed through the power spectral density of the covariance, which is given by the Fourier transformation (Press et al., 1992) of the covariance function. Power spectral density is defined by the integral:

$$\widetilde{c}_Z(\mathbf{k}) = \int d\mathbf{r} \exp(-i\mathbf{k}\cdot\mathbf{r}) c_Z(\mathbf{r}), \tag{1.7}$$

where $\mathbf{r}$ is the distance vector between two points, $\int d\mathbf{r} = \int dx \int dy$, and $\mathbf{k}$ is the vector of spatial frequency (wavevector). Function $c_Z(\mathbf{r})$ is an accepted covariance function if the three following conditions are applicable: (1) the power spectral density exists, i.e., if the Fourier transformation of the function exists, (2) the power spectral density is nonnegative, and (3) the integral of the power spectral density throughout the range of frequencies is bounded (i.e., if the variance exists). In practice, to determine if a function is an acceptable covariance, the Fourier transform of the function needs to be calculated (Cressie, 1993).

2.4 Statistical Homogeneity

Assumptions that impose constraints on the properties of an RF can lead to a more efficient geostatistical analysis. The most widely used simplifying assumption is statistical homogeneity, which is an extension of the classical definition of homogeneity. A given property is homogeneous if the corresponding variable is constant in space. On the contrary, an RF is statistically homogeneous if the mean value is constant, $m'_Z(\mathbf{s}) = m'_Z$, the covariance function is defined and depends only on the distance vector $\mathbf{r} = \mathbf{s}_1 - \mathbf{s}_2$ between two points $c_Z(\mathbf{s}_1, \mathbf{s}_2) = c_Z(\mathbf{r})$, and the variance of the field is also constant. These conditions also define second-order stationarity.

These conditions define the statistical homogeneity in a weak sense. An RF is statistically homogeneous in a strong sense when the multidimensional PDF for N points (where N is any positive integer number) remains unchanged by transformations that alter the location of the points without altering the distances between them. Therefore the concept of statistical homogeneity is that the statistical properties of a random field do not depend on the spatial coordinates of the points, hence the reference system. Practically, statistical homogeneity implies that there are no systematic trends, so the change of values in the field can be attributed to fluctuations around a constant level equal to the mean value (Goovaerts, 1997).

2.5 Statistical Isotropy

Another property that can be useful in geostatistical analysis of an RF is statistical isotropy. A field is statistically isotropic if it is statistically homogeneous and at the same time the covariance function depends on the distance (Euclidean distance), but not on the direction of the distance vector $\mathbf{r}$. This is important from a practical point of view because it helps in the identification of spatial dependence. If a covariance function is statistically isotropic it is by definition statistically homogeneous, but not vice versa.

In the case of statistically isotropic fields the two most important parameters that determine the very basic features of the covariance function are variance $\sigma_Z^2 = c_Z(0)$ and correlation length ξ. Variance is a measure of the width of the fluctuations in the field. The correlation length defines the interval in which there is interdependence, which defines the distance within which the field value at one point affects the value at another point (Christakos, 1991).

2.6 Spatial Dependence

There are several ways to measure spatial dependence. Two of the most commonly used are the semivariogram and the correlation function. Both functions describe the dependence between two points in the statistical sense as both functions refer to pairs of points, so their value depends on the distance between these points. The term, in the statistical sense, means that the described dependence emerges as a mean value from a large number of pairs and not a single pair of points (Hristopulos, 2008). Correlation function for a random field is equal to the ratio of the covariance function to the variance, while the semivariogram of a random field is defined by the equation:

$$\gamma_Z(\mathbf{s},\mathbf{r}) = \frac{1}{2}E\left\{[Z(\mathbf{s}+\mathbf{r}) - Z(\mathbf{s})]^2\right\}. \tag{1.8}$$

The semivariogram is defined in relation to a pair of points using the mean squared difference: $\delta Z(\mathbf{s};\mathbf{r}) \equiv Z(\mathbf{s}+\mathbf{r}) - Z(\mathbf{s})$. The difference field $\delta Z(\mathbf{s};\mathbf{r})$ is called distance step $\mathbf{r}$. If the field $Z(\mathbf{s})$ is statistically homogeneous the semivariogram is directly related to the covariance function by the equation:

$$\gamma_Z(\mathbf{r}) = \sigma_Z^2 - c_Z(\mathbf{r}). \tag{1.9}$$

For statistically homogeneous fields, the semivariogram contains the same information as the covariance function. If the difference $\delta Z(\mathbf{s};\mathbf{r})$ is statistically homogeneous, the random field $Z(\mathbf{s})$ is called the field with statistically homogeneous differences. In this case the semivariogram $\gamma_Z(\mathbf{r})$ depends solely on the distance $\mathbf{r}$ between the points and this is a result of statistical homogeneity of the field differences. If the field $Z(\mathbf{s})$ is statistically homogeneous then the same applies for the difference $\delta Z(\mathbf{s};\mathbf{r})$, but the opposite is not necessarily true (Deutsch and Journel, 1992).

The parameters of the semivariogram determine the spatial dependence of the field values at two neighboring points. From the definition of the semivariogram, using the mean square of the differences, it is shown that the semivariogram is semipositively defined, $\gamma_Z(\mathbf{r}) \geq 0$. However, the reverse is not always the case, because a semipositive defined function is not necessarily an admissible semivariogram.

In the case of a statistically homogeneous field, if the spatial dependence is isotropic, the semivariogram is determined by two parameters: the *sill* and the *correlation length.* The value of the semivariogram for long distances $\mathbf{r}$ tends asymptotically to a sill equal to the variance σ_Z^2 of the random field. This property is based on $\gamma_Z(\mathbf{r}) = \sigma_Z^2 - c_Z(\mathbf{r})$ and the fact that at large distances the value of the covariance function tends toward zero. The presence of important large distance trends means that the assumption of statistical homogeneity is not valid. Then the semivariogram does not converge toward a balance value when the distance tends toward infinity (Olea, 1999).

If correlation characteristics vary in different directions in space then the dependence is anisotropic. The two main types of anisotropy that are encountered in practice are geometrical and zone anisotropy. Geometrical anisotropy refers to cases when the semivariogram sill is independent of the direction, but the velocity approaching the sill depends on the direction (Hohn, 1999). In this case the semivariogram is expressed as function $\gamma_Z\left(\frac{r_1}{\xi_1},\cdots,\frac{r_d}{\xi_d}\right)$ of nondimensional distances $\frac{r_1}{\xi_1},\cdots,\frac{r_d}{\xi_d}$, where $\xi_1,\ldots,\xi_d$ are the correlation lengths in the corresponding directions.

Zone anisotropy refers to the case where the sill depends on the spatial direction. Then the semivariogram can be expressed as the sum of the resultant $\gamma_Z(\mathbf{r}) = \gamma_{Z,1}(r) + \gamma_{Z,2}(\widehat{\mathbf{r}})$. In this, the equation $\gamma_{Z,1}(r)$, where $r = \|\mathbf{r}\|$, describes an isotropic dependence, while $\gamma_{Z,2}(\widehat{\mathbf{r}})$ describes the anisotropic dependence between the sill and the direction of the unit vector $\widehat{\mathbf{r}}$.

In the case of geometrical anisotropy, more than one correlation length is required, $\xi_1,\ldots,\xi_d$. Some of them, but not all, may be equal to each other. Therefore additional parameters are required for the determination of the semivariogram's anisotropy. In a two-dimensional system, ξ_x and ξ_y correspond to the correlation lengths along the main axis; the anisotropy parameters are: (1) the ratio $\rho_{y/x} \equiv \xi_y/\xi_x$, which is called the anisotropy ratio, and (2) the orientation angle, which defines the orientation of the main anisotropy axis in relation to the Cartesian coordinate system.

To understand the meaning of the orientation angle, the ellipse is defined as the geometrical location of points (r_x, r_y), where the value of the semivariogram is constant. The elliptical shape is used since this happens for different semivariogram models, such as exponential and Gaussian anisotropic semivariograms. The orientation angle is the angle between the KA1 axis of the ellipsis and the horizontal axis of the coordinate system (Fig. 1.1; Hristopulos, 2008).

The semivariogram generally increases, but not necessarily linearly, with the distance between the points, while the correlation function decreases. This is because the correlation function describes the dependence between the field values in two different points in space and their dependence decreases in larger distances. On the contrary, the semivariogram measures the difference between field values as a function of their distance. Therefore semivariogram values increase when the distance increases. For statistically homogeneous fields, the two functions are equivalent, which means that

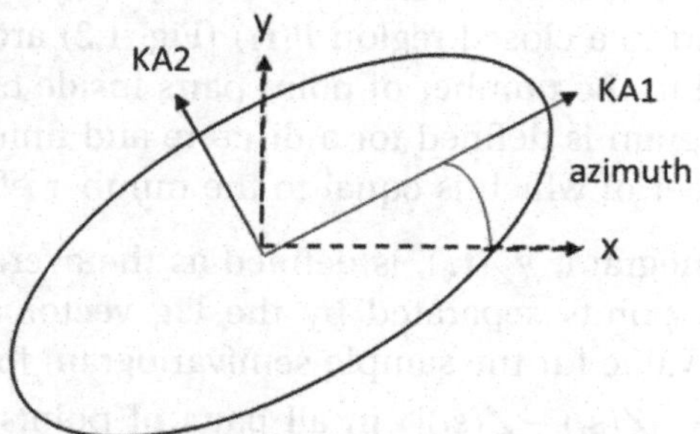

FIGURE 1.1 Presentation of the main axis system (KA1, KA2) in relation to the coordinate system *x,y*. The ellipsis corresponds to the semivariogram direction. *After Hristopulos, D. T., 2008. Applied Geostatistics-Cource Notes (In Greek), vol. 200. Chania, Crete, Greece: Technical University of Crete.*

they have the same information in different forms. However, there are cases of random fields where the semivariogram is a function of the distance between two points only, while the correlation function depends both on the distance and the specific location of the points in space (Deutsch and Journel, 1992).

Widely used semivariogram models, which can also be used in practical applications, are the exponential, Gaussian, spherical, power law, and nugget effect. The exponential model characterizes distributions with sharp spatial variations, opposite to the Gaussian model, which characterize smoother variations. The power law model corresponds to dependence with a long-distance spatial range, and the nugget effect corresponds to variations that take place at distances smaller than the resolution of the sample allows. Another way of determining the spatial dependence of a random field is the method of the Spartan variogram family (Hristopulos, 2003b).

2.7 Semivariogram Estimation

The main mathematical tool in geostatistical modeling is the semivariogram, which expresses the spatial dependence between neighboring observations. In the case of geographical distributions and distribution of environmental variables, where the available data are limited to a sole sample, an attempt is made to determine an estimation of the real semivariogram through the sample. This estimation is called sampled or experimental semivariogram and is calculated based on the values of the sample. The Matheron method-of-moments estimator of the semivariogram is given by (Isaaks and Srivastava, 1989; Deutsch and Journel, 1992):

$$\widehat{\gamma}_Z(\mathbf{r}_k) = \frac{1}{2N(\mathbf{r}_k)} \sum_{i,j=1}^{N(\mathbf{r}_k)} \left\{ [Z(\mathbf{s}_i) - Z(\mathbf{s}_j)]^2 \right\} \vartheta_{i,j}(\mathbf{r}_k), \quad (k = 1, \ldots, N_c), \tag{1.10}$$

$$\vartheta_{ij}(\mathbf{r}_k) = \left\langle \begin{matrix} 1, \mathbf{s}_i - \mathbf{s}_j \in B(\mathbf{r}_k) \\ 0, \text{ otherwise} \end{matrix} \right\rangle.$$

- The class function $\vartheta_{i\,j}(\mathbf{r}_k)$ defines different classes of distance vectors, choosing the vectors that correspond to a closed region $B(\mathbf{r}_k)$ (Fig. 1.2) around vector $\mathbf{r}_k$.
- Variable $N(\mathbf{r}_k)$ is equal to the number of point pairs inside class $B(\mathbf{r}_k)$.
- The sample semivariogram is defined for a discrete and finite set of distances $\mathbf{r}_k$, $(k = 1,\ldots,N_c)$ the number of which is equal to the number of classes N_c.

The empirical semivariogram, $\widehat{\gamma}_Z(\mathbf{r}_k)$, is defined as the average square difference of the field values between points separated by the lag vector $\mathbf{r}_k$. More precisely, this calculation determines a value for the sample semivariogram for every $\mathbf{r}_k$, based on the mean value of differences $[Z(\mathbf{s}_i) - Z(\mathbf{s}_j)]^2$ in all pairs of points, the distance vector of which belongs in the $B(\mathbf{r}_k)$ region. $\widehat{\gamma}_Z(\mathbf{r}_k)$ is a good estimator of the real $\gamma_Z(\mathbf{r}_k)$ when the mean value of differences in the $\mathbf{r}_k$ class approaches with accuracy the mean value

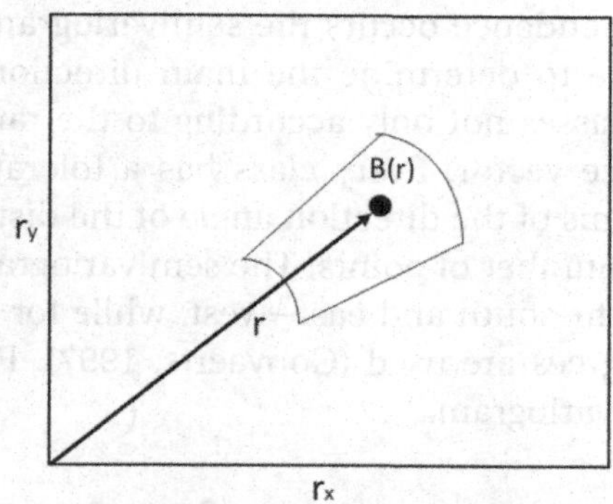

FIGURE 1.2 Schematic figure of the region *B*(**r**) around the distance vector (Hristopulos, 2008).

$E[Z(\mathbf{s})-Z(\mathbf{s}+\mathbf{r}_k)]^2$. The latter is true when the ergodic assumption applies, which allows the switch between the stochastic and the sample mean. In semivariogram calculation the ergodic property is valid when the following conditions occur: the field of differences $Z(\mathbf{s})-Z(\mathbf{s}+\mathbf{r}_k)$ is statistically homogeneous, the number of pairs in each class is large enough so the sample mean of the square difference is determined with good statistical accuracy, and the number of classes is large enough so the dense approach of the semivariogram variations as a function of the distance is allowed. After the sampled semivariogram is calculated, it is adjusted to a theoretical model, which allows the calculation of the semivariogram in every distance. This can be achieved using, e.g., the least squares method, from which the optimal values for parameters ξ and σ_Z^2 of the theoretical model can be calculated. The variable ξ denotes the characteristic (correlation) length and σ_Z^2 the variance (sill) (Olea, 1999).

The theoretical model is needed for the estimation of field values in points where measurements are not available. Next, to accept the semivariogram and use it for geostatistical analysis, it is tested according to the semivariogram acceptance conditions. A semivariogram is acceptable if it is conditionally negative definite. This means that for any linear coefficients λ_α that satisfy the equation $\sum_{a=1}^{N} \lambda_\alpha = 0$ the following inequality must apply:

$$-\sum_{a=1}^{N}\sum_{\beta=1}^{N}\lambda_\alpha\lambda_\beta\gamma_Z(\mathrm{s}_\alpha-\mathrm{s}_b)\geq 0, \tag{1.11}$$

for any positive integer N. For a spatial homogeneous random field it is simpler to check the acceptance of a semivariogram or covariance model using the function $\sigma_Z^2-\gamma_Z(\mathbf{r})$. If the function $\gamma_Z(\mathbf{r})$ describes an acceptable semivariogram, then the function $c_Z(\mathbf{r})=\sigma_Z^2-\gamma_Z(\mathbf{r})$ is an acceptable covariance function and vice versa (Deutsch and Journel, 1992).

If anisotropic spatial dependence occurs the semivariogram should be calculated in different directions in space to determine the main direction of the anisotropy. This requires the definition of classes not only according to the range but also according to the direction of the distance vector. Every class has a tolerance ($2\delta r$) in terms of the range, as well as ($2\delta\phi$) in terms of the direction angle of the distance vector, so that every class includes an adequate number of points. The semivariogram is usually calculated in terms of the directions north–south and east–west, while for the angular tolerance the values 5, 10, 20, and 45 degrees are used (Goovaerts, 1997). Fig. 1.3 presents the characteristics of a typical semivariogram.

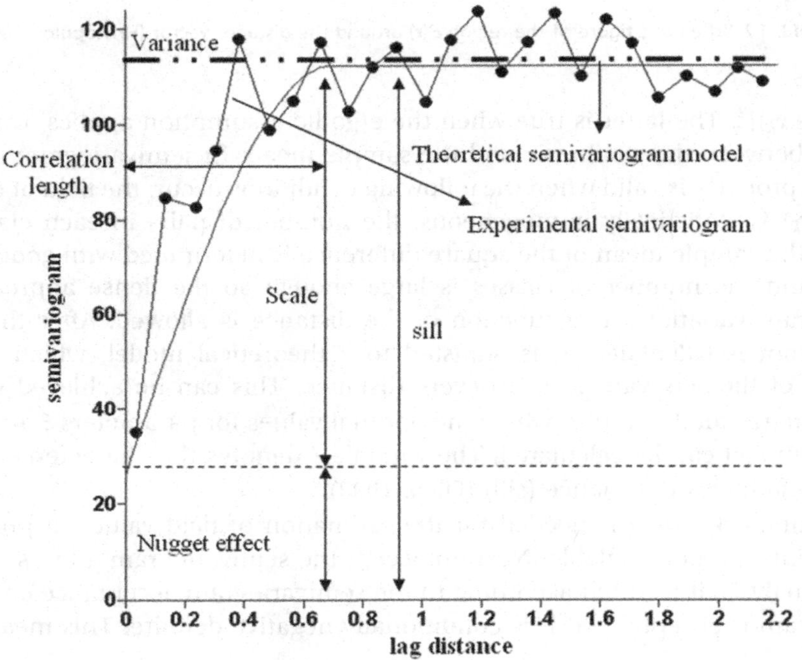

FIGURE 1.3 Presentation of typical semivariogram characteristics.

- The nugget effect quantifies the variance of the sampling error, as well as the small-scale variance, e.g., the spatial variance in distances smaller than the distances between sampling points.
- The sill is the value that approaches asymptotically the experimental semivariogram.
- Scale is the difference between the sill and the nugget effect, and declares the variance of the correlated fluctuations.
- The correlation length is the distance in which the semivariogram almost (e.g., 95%–97%) reaches the sill value.
- Variance is the mean squared deviation of every value of the sample from the mean value and is denoted with the horizontal dashed line in the figure.
- The experimental semivariogram represents the classes of pairs along with the corresponding sampled values of the semivariogram.
- The theoretical semivariogram model is a continuous theoretical line that is fitted to the experimental semivariogram.

If there are no distinct anisotropies, the omnidirectional empirical semivariogram $\widehat{\gamma}_Z(\mathbf{r})$, $\mathbf{r} = \mathbf{r}_k$, is estimated and then fitted to a theoretical model function $\gamma_Z(\mathbf{r})$ (Deutsch and Journel, 1992; Kitanidis, 1997).

2.8 Semivariogram Models

The following classical theoretical semivariogram models include the spherical, Gaussian, exponential, power law, and linear functions (Goovaerts, 1997; Lantuejoul, 2002); σ_z^2 is the variance, $|\mathbf{r}|$ is the Euclidean norm of the lag vector $\mathbf{r}$, and ξ is the characteristic length:

$$\text{Exponential}: \gamma_Z(\mathbf{r}) = \sigma_Z^2\left[1 - \exp\left(-\frac{|\mathbf{r}|}{\xi}\right)\right] \tag{1.12}$$

$$\text{Gaussian}: \gamma_Z(\mathbf{r}) = \sigma_Z^2\left[1 - \exp\left(-\frac{\mathbf{r}^2}{\xi^2}\right)\right] \tag{1.13}$$

$$\begin{gathered}\text{Spherical}: \gamma_Z(\mathbf{r}) = \sigma_Z^2\left[1.5|\mathbf{r}|/\xi - 0.5(|\mathbf{r}|/\xi)^3\right]\theta(\xi - |\mathbf{r}|) \\ \text{if } \ \xi - |\mathbf{r}| < 0, \theta = 0, \text{else if } \ \xi - |\mathbf{r}| > 0, \ \ \theta = 1\end{gathered} \tag{1.14}$$

$$\text{Power law}: \gamma_Z(\mathbf{r}) = c|\mathbf{r}|^{2\mathrm{H}}, 0 < \mathrm{H} < 1 \tag{1.15}$$

where c is the coefficient and H is the Hurst exponent.

$$\text{Linear}: \gamma_Z(\mathbf{r}) = c|\mathbf{r}| \tag{1.16}$$

These equations define the isotropic versions of the models. These involve at most two parameters, i.e., the variance and correlation length for exponential, Gaussian, and spherical models, c and H for the power law model, and c for the linear model. We now review two semivariogram models that offer increased parameter flexibility.

2.9 Matérn Model

This covariance family includes, in addition to the variance and correlation length, a smoothness parameter ν, which controls the continuity and differentiability of the random field, and thus also the short-distance behavior of $\gamma(\mathbf{r})$, which has greater impact on interpolation than medium-to-large distance dependence. The Matérn semivariogram model (Matérn, 1986; Stein, 1999; Pardo-Iguzquiza and Chica-Olmo, 2008) is defined as:

$$\gamma_Z(\mathbf{r}) = \sigma_Z^2\left\{1 - \frac{2^{1-\nu}}{\Gamma(\nu)}\left(\frac{|\mathbf{r}|}{\xi}\right)^{\nu} K_\nu\left(\frac{|\mathbf{r}|}{\xi}\right)\right\}, \tag{1.17}$$

where $\sigma_z^2 > 0$ is the variance, $\xi > 0$ is the characteristic length, $\nu > 0$ is the smoothness parameter, $\Gamma(\cdot)$ is the gamma function, $K_\nu(\cdot)$ is the modified Bessel function of the second kind of order ν, and $|\mathbf{r}|$ is the Euclidean norm of vector $\mathbf{r}$. For $\nu = 0.5$ the exponential model is recovered, whereas the Gaussian model is obtained at the limit as ν

tends to infinity. The case $\nu = 1$ was introduced by Whittle (1954). The Matérn model has been applied to different research fields, including hydrology, e.g., Rodriguez-Iturbe and Mejia (1974) and Zimmermann et al. (2008).

2.10 Spartan Model

Spartan spatial random fields (SSRFs) are a recently proposed geostatistical model (Hristopulos 2002, 2003b) with applications in environmental risk assessment (Elogne et al., 2008) and atmospheric environment (Žukovič and Hristopulos, 2008). SSRFs are generalized Gibbs random fields, equipped with a coarse-graining kernel that acts as a low-pass filter for the fluctuations. The term Spartan indicates parametrically compact model families that involve a small number of parameters. These random fields are defined by means of physically motivated spatial interactions between the field values.

In general, an SRF $Z(\mathbf{s})$ representing the measurements can be expressed as:

$$Z(\mathbf{s}) = Z'(\mathbf{s}) + m_Z(\mathbf{s}) + \mathrm{e}(\mathbf{s}), \tag{1.18}$$

where $\mathrm{e}(\mathbf{s})$ is a zero-mean measurement noise process, assumed to be homogeneous over the domain of interest, $Z'(\mathbf{s})$ is a correlated fluctuation SRF, and $m_Z(\mathbf{s})$ is a deterministic trend function. The trend is a nonstationary component representing large-scale, deterministic variations, which presumably correspond to the ensemble average of the SRF:

$$m_Z(\mathbf{s}) = E[Z'(\mathbf{s})]. \tag{1.19}$$

SSRFs are determined from a PDF in terms of an SRF $Z'(\mathbf{s})$. The PDF contains information for spatial dependence. In general, the PDF SSRF can be expressed with the following equation:

$$f_x[Z'(\mathbf{s})] = Z^{-1} \exp\{-H[Z'(\mathbf{s})]\}, \tag{1.20}$$

$$Z = \sum_{z'(\mathbf{s})} \exp\{-H[Z'(\mathbf{s})]\}, \tag{1.21}$$

which is a normalization constant that ensures the basic theorem of probability (i.e., that the *sum of probabilities* of an SRF is equal to 1). $H[Z'(\mathbf{s})]$ is an energy functional of spatial dependence, which expresses the interdependence of SRF data values $Z'(\mathbf{s})$ between different locations. Therefore SSRFs belong in the family of Gibbs random fields (Hristopulos, 2003b). The Gibbs property stems from the fact that the joint PDF of SSRFs is expressed in terms of an energy functional, i.e., $H[Z'(\mathbf{s})]$. Use of an energy functional containing terms with a clear physical interpretation permits inference of the model parameters based on matching respective sample constraints with their ensemble values (Hristopulos and Elogne, 2007). Thus the spatial continuity properties can be determined without estimation of the experimental variogram.

The SSRFs provide a new class of generalized covariance functions that are by construction positive definite for an explicitly specified range of parameter values (Hristopulos, 2003b; Hristopulos and Elogne, 2007). Fluctuation-gradient-curvature

(FGC) SSRF models have an energy functional that involves the squares of the fluctuations, the gradient, and the curvature of the field, see Eq. (1.22). This class provides covariance functions with four parameters that give considerable flexibility. The SSRF covariance functions can be used for spatial interpolation with the classical kriging estimators as well as with new spatial predictors (Elogne et al., 2008; Hristopulos and Elogne, 2009).

Herein we use this new class of covariance function for the first time in groundwater hydrology in association with kriging for spatial interpolation of the groundwater level. For kriging applications, the estimation of the spatial dependence structure (semivariogram or covariance function) is a crucial step.

The isotropic FGC-SSRF functional is given by the following equation:

$$H[Z'(\mathbf{s})] = \frac{1}{2\eta_0 \xi^d} \int d\mathbf{s}\left[\{Z'(\mathbf{s})\}^2 + \eta_1 \xi^2 \{\nabla Z'(\mathbf{s})\}^2 + \xi^4 \{\nabla^2 Z'(\mathbf{s})\}^2\right]. \tag{1.22}$$

The FGC model involves the parameters η_0, η_1, ξ, and k_m. The scale coefficient η_0 determines the overall scale of the variance; the scale factor is proportional to the square of the regionalized variable, i.e., the groundwater level, and assumes the variable's units. The shape coefficient η_1 is dimensionless and determines the shape of the covariance function in connection with ξ and k_m. The characteristic length ξ has dimensions of length and determines the range of spatial dependence. Finally, the wavevector k_m has units of inverse length and determines the bandwidth of the covariance spectral density. If the latter is band limited, k_m represents the band cutoff and is related to the resolution length scale by means of $k_m \lambda \approx 1$.

The SSRF covariance models derived from the foregoing energy functional are determined by the parameters $\boldsymbol{\theta} = (\boldsymbol{\eta_0}, \boldsymbol{\eta_1}, \boldsymbol{\xi}, \mathbf{k_m})$. Spartan covariance and semivariogram functions were introduced in Hristopulos (2003b) and have been applied to various environmental datasets. Herein, we apply this family of functions for the first time in hydrological data.

The SSRF family includes four-parameter functions (Hristopulos, 2003b; Hristopulos and Elogne, 2007). The Spartan covariance in any dimension d is expressed using the spectral representation as follows:

$$C_Z(\mathbf{r};\boldsymbol{\theta}) = \frac{\eta_0 \xi \|\mathbf{r}\|^{1-d/2}}{(2\pi)^{d/2}} \int_0^{k_m} d\omega \, \frac{\omega^{d/2} J_{d/2-1}(\|\mathbf{r}\|\omega)}{1+\eta_1(\omega\xi)^2+(\omega\xi)^4}, \tag{1.23}$$

where $J_{d/2-1}(x)$ is the Bessel function of the first kind of order zero and $\boldsymbol{\theta} = (\eta_0, \eta_1, \xi, k_m)$ are the model parameters. The Spartan semivariogram is given by $\gamma_Z(\mathbf{r};\ \boldsymbol{\theta}) = C_Z(0;\ \boldsymbol{\theta}) - C_Z(\mathbf{r};\ \boldsymbol{\theta})$. The scale parameter η_0 determines the variance, ξ is the characteristic length, k_m represents the wavenumber cutoff (band limit in Fourier space), and the dimensionless stiffness coefficient η_1 determines the shape of the covariance function in connection with k_m and ξ (Elogne et al., 2008). In $d = 1{,}3$ explicit expressions for the Spartan covariance are possible at the asymptotic limit $k_m \to \infty$ (Hristopulos and Elogne, 2007).

The Spartan covariance function of Eq. (1.23) in $d = 3$ dimensions is expressed as follows (Varouchakis and Hristopulos, 2013):

$$C_Z(\mathbf{h};\boldsymbol{\theta}) = \begin{cases} \dfrac{\eta_0 e^{-h\beta_2}}{2\pi\sqrt{|\eta_1^2-4|}}\left[\dfrac{\sin(h\beta_1)}{h\beta_1}\right], & \text{for} |\eta_1| < 2, \sigma_z^2 = \dfrac{\eta_0}{2\pi\sqrt{|\eta_1^2-4|}} \\ \dfrac{\eta_0 e^{-h}}{8\pi}, & \text{for } \eta_1 = 2, \sigma_z^2 = \dfrac{\eta_0}{8\pi} \\ \dfrac{\eta_0(e^{-h\omega_1} - e^{-h\omega_2})}{4\pi(\omega_2-\omega_1)h\sqrt{|\eta_1^2-4|}}, & \text{for } \eta_1 > 2, \sigma_z^2 = \dfrac{\eta_0}{4\pi\sqrt{|\eta_1^2-4|}} \end{cases}, \tag{1.24}$$

$$\omega_{1,2} = (|\eta_1 \mp \Delta|/2)^{1/2}, \tag{1.25}$$

$$\beta_{1,2} = |2 \mp \eta_1|^{1/2}/2. \tag{1.26}$$

In the foregoing, $\Delta = |\eta_1^2 - 4|^{1/2}$, $\omega_{1,2}$, and β_2 are dimensionless damping coefficients, β_1 is a dimensionless wavenumber, ξ is a characteristic length, $\|\mathbf{h}\| = \|\mathbf{r}\|/\xi$ is the normalized lag vector, $h = |\mathbf{h}|$ is the Euclidean norm, and σ_z^2 is the variance. The exponential covariance is recovered for $\eta_1 = 2$, while for $|\eta_1| < 2$ the product of the exponential and hole-effect model is obtained. A covariance function that is permissible in *three* spatial dimensions is also permissible in two dimensions (Christakos, 1991). Hence Eq. (1.24) can be used in two dimensions, albeit it does not correspond to the FGC-SSRF two-dimensional covariance (Hristopulos and Elogne, 2007).

2.11 Parameter Inference

The Spartan parameters can be estimated by fitting the SSRF semivariogram to the empirical semivariogram estimator. A different approach is based on the modified method of moments, in which stochastic constraints are matched with corresponding sample constraints (Elogne et al., 2008; Žukovič and Hristopulos 2008, 2009). The constraints are motivated by the terms in the energy functional (Eq. 1.22); the square of the fluctuations, the square gradient, and the square curvature are used to construct both the sample and the stochastic constraints. The latter approach is not investigated herein because the focus of these notes is on kriging interpolation techniques.

There is no universally accepted method for fitting the empirical semivariogram to a theoretical model. For each of the foregoing theoretical models we determine the optimal semivariogram parameters using the least squares method. Methods used include least squares fits, weighted least squares, generalized least squares, maximum likelihood, and even empirical "fitting by eye" (Wackernagel, 2003; Olea, 2006). We implemented least squares fitting by means of the fminsearch Matlab function, which is based on the Nelder–Mead minimization algorithm (Press et al., 1992). The selection of the "optimal semivariogram model" is based on the results of leave-one-out cross-validation (see Section 2.5).

2.12 Spatial Estimation

Determination of the spatial dependence, as well as the trend and fluctuations of the field values, leads in two basic procedures of geostatistics: spatial estimation and simulation. Both procedures help in the representation of a random field in points where inexact values exist, based on available information (e.g., measurements in neighboring points, hydrogeological data). The available information is used to impose statistical limitations. Using statistical spatial dependence patterns (semivariograms) the unknown values are defined based on their correlation. The repetitiveness of this procedure in all points of the computational grid allows mapping of an entire area (Hristopulos, 2008).

The simulation process aims to create many of the possible states of the field, which are in accordance with existing statistical restrictions derived by the experimental sample, e.g., simulated states with the same mean value, standard deviation, and semivariogram with the one calculated using samples. Therefore the simulation aims to create many alternative scenarios, which are possible based on existing measurements (Hristopulos, 2008).

The term spatial and/or temporal estimate includes all the mathematical procedures that allow the calculation of field values where measurements of a property do not exist. The estimate can be local, if it is referred to a point in space–time, or global, if it aims to calculate a characteristic value that describes an entire region. The spatial and/or temporal estimate of a field presupposes the existence of spatial and/or temporal dependence, so that the field value at each point is "influenced" by the neighboring field values. This interdependence allows estimation of a variable where measurements do not exist based on the neighboring measured points. In many cases, the final objective is to estimate the field over a set of points instead of a single one. Various methods of spatial estimation (interpolation) exist that are based on similar principles. The main idea is that the value at the estimation point is given by a linear or nonlinear combination of the neighboring values. The estimate results from the optimization of a statistical measure, e.g., maximization of probability or minimization of the mean square estimation error. The most popular methods are based on linear interpolation in conjunction with the minimization of the mean square estimate error. This set of methods is known as "kriging" (Goovaerts, 1997; Kitanidis, 1997; Hristopulos, 2003a).

The need of variables estimation at points where no measurements are available is not new. Statistical scientists, mining engineers, oil engineers, hydrologists, and geologists who dealt with the problem developed the science of geostatistics. Application areas of geostatistics nowadays include: the analysis of ore deposits (e.g., estimate of extent, depth, and quantification of total content), e.g., Journel and Huijbregts (1978) and De-Vitry et al. (2010), oceanography (mapping of the ocean bed, wave height analysis), e.g., Özger and Şen (2007), the morphological analysis of natural and technological nonhomogeneous (e.g., porous) materials, e.g., Sahimi (2011), the mapping of pollutant concentrations in various environmental means (air, subsoil, surface-underground water resources), e.g., Goovaerts (1997) and Webster and Oliver (2001), topographic analysis and geographic information systems, e.g., Burrough (2001), the spatiotemporal analysis of rainfall data and of rainfalls in regions with insufficient monitoring stations, e.g., Ly et al. (2011), the determination of geological and hydrogeological variables (e.g., subsoil type, hydraulic conductivity, porosity, storativity, evapotranspiration), e.g., Kitanidis (1997) and Hengl (2007), environmental and human health risk assessment (e.g., estimate of pollutant concentration, determination of probabilities of exceeding the critical limits), e.g., Goovaerts (1997) and Christakos and Hristopulos (1998), and the spatial and/or temporal estimation of the hydraulic head of aquifers, e.g., Ahmed (2007).

3. Spatial Interpolation

Geostatistics is based on the work of Kolmogorov (1941) in atmospheric turbulence. He used the structure function (equivalent to the variogram) to represent spatial correlations and to develop optimal interpolation. Later, Matérn developed the family of spatial covariance functions that bears his name (Matérn, 1986). His functions are equivalent to those developed by Jowett (1955). The geostatistical method called kriging, the most applied geostatistical method to date, was introduced and established by Krige (1951, 1966) and Matheron (1963) for applications in mining engineering. Since then, kriging has been applied to several other fields of research, such as geology (Davis, 1973; Journel and Huijbregts, 1978), petroleum engineering (Hohn, 1999), hydrogeology (Kitanidis, 1997), hydrology, meteorology, and soil science (Webster and Oliver, 2001; Atkinson and Lloyd, 2010). The first application of kriging in groundwater hydrology was by Delhomme (1974). Since then, many studies applied kriging to the interpolation of groundwater levels, e.g., Delhomme (1978), Gambolati and Volpi (1979a,b), Sophocleous et al. (1982), Aboufirassi and Marino (1983), Pucci and Murashige (1987), Hoeksema et al. (1989), Desbarats et al. (2002), Ahmadi and Sedghamiz (2007), Kumar (2007), Ahmadi and Sedghamiz (2008), Rivest et al. (2008), and Nikroo et al. (2009).

In general, interpolation methods routinely used for groundwater level mapping include deterministic methods such as inverse distance weighting (IDW) (Gambolati and Volpi, 1979b; Philip and Watson, 1986; Rouhani, 1986; Buchanan and Triantafilis, 2009; Sun et al., 2009) and stochastic methods such as ordinary kriging (OK) and universal kriging (UK). Such methods are incorporated in various commercial software packages, e.g., mapping software: Arc-View (geographic information systems) and Surfer mapping system (Golden software) and groundwater modeling software: Visual Modflow, Princeton Transport Code, and Feflow subsurface flow model.

Deterministic interpolation methods use closed-form mathematical formulas (IDW) or the solution of a linear system of equations (minimum curvature [MC]) to interpolate the data. The weights assigned to each sample value depend only on the distance between the sample point and the location of the interpolated point. Deterministic methods are categorized as global and local: global methods use the entire dataset for prediction at each point, while local methods use data in a neighborhood around the interpolation point. Deterministic methods can be either exact or inexact interpolators (Webster and Oliver, 2001). Finally, they do not generate measures of estimate uncertainty.

Stochastic methods employ the spatial correlations between values at neighboring points. The most widely used stochastic method is kriging (Krige, 1951; Matheron, 1963, 1971). The kriging methodology comprises a family of interpolators. The interpolators most commonly used in hydrosciences are OK and UK. A recently proposed variation of the kriging algorithm is kriging with Delaunay triangulation (DK) (Hessami et al., 2001).

Kriging is characterized as the best linear unbiased estimator. The kriging estimator is a weighted linear function of the data. The linear weights follow from the unbiasedness constraint (i.e., zero mean estimation error) and the minimum square error condition. The resulting system of linear equations is solved to determine the estimator's weights. The coefficients of the equations depend on the model semivariogram, which is obtained by fitting the empirical semivariogram to theoretical models or by means of the maximum likelihood estimation method (Kitanidis, 1997; Ahmed, 2007). The semivariogram measures the degree of spatial correlation as a function of distance and/or direction between data points. The semivariogram determines the kriging weights and therefore controls the quality of the estimates (Mouser et al., 2005; Ahmed, 2007). If the semivariogram is perfectly known, kriging is the best linear unbiased estimator. An advantage of kriging compared to deterministic approaches is that it allows the estimation of the interpolation error at unmeasured points (Deutsch and Journel, 1992). In addition, in the absence of a nugget (e.g., measurement error), kriging is an exact interpolator at measurement points (Delhomme, 1974; Ahmed, 2007). Optimal kriging results are obtained if the probability distribution of the data is normal and stationary in space (spatially homogeneous). Kriging is computationally intensive when applied to large datasets (Webster and Oliver, 2001), but the computational complexity is not a problem for sparsely sampled areas.

OK bases its estimates on unsampled locations only on the sampled primary variable. OK interpolation is widely used to determine the spatial variability of groundwater levels in hydrological basins, e.g., OK was also used to predict the piezometric head in West Texas and New Mexico based on implementing clustered piezometric data (Abedini et al., 2008). In addition, the design, evaluation, and optimization of groundwater level monitoring networks were performed by applying OK. Evaluation of the performance and interpolation errors of OK in the estimation of water level elevation can be achieved by means of leave-one-out cross-validation (Olea, 1999).

OK is not optimal for nonstationary data. The use of a linear drift term improves the accuracy of the interpolated head field if a regional gradient is present (Delhomme, 1978; Aboufirassi and Marino, 1983). UK also has been used to estimate the groundwater level, e.g., near extraction or injection wells a point logarithmic component is added to the drift to account for the drawdown caused by the pumping well. This approach is applicable if analytical solutions for the aquifer response are available. Auxiliary information can be included in the interpolation as a drift term, usually modeled by polynomial functions of the space coordinates, rainfall, or surface elevation based on a digital elevation model (DEM). The use of auxiliary variables in general improves the accuracy of kriging estimation. Easily measurable secondary variables can also reduce the number of "expensive" observations (Knotters et al., 1995). The auxiliary information can be incorporated using the cokriging (CoK) method, which utilizes secondary variables in the covariance structure. Various researchers used CoK with ground surface elevation as a secondary variable to construct groundwater level maps that improved the OK predictions. The main disadvantage of CoK is the need to model coregionalization, which requires the inference of direct and cross-covariance functions (Journel and Huijbregts, 1978). CoK also becomes cumbersome and time consuming if many secondary variables are involved (Deutsch and Journel, 1992).

Alternatively, residual kriging (RK) and kriging with external drift (KED), originally described and applied in hydrological problems, embody secondary information in the drift term. KED and RK are practically equivalent but differ in the methodological steps used (Hengl et al., 2003; Hengl, 2007). RK is also known as regression kriging and was developed and applied in the hydrosciences by Delhomme (1974, 1978) and Ahmed and De Marsily (1987). Odeh et al. called it regression kriging, while Goovaerts (1999) used the term kriging after detrending (Hengl et al., 2003).

KED assumes that the expectation of the primary variable is a linear combination of secondary variables (Deutsch and Journel, 1992; Wackernagel, 2003), while OK assumes the expectation to be constant (Rivest et al., 2008). In the case of KED (it has similar methodology to UK), the kriging covariance matrix of residuals is extended with the auxiliary predictors (Kitanidis, 1997; Webster and Oliver, 2001). KED was applied for the interpolation of water table elevation by various researchers. Beven and Kirkby (1979) expressed the water table depth as a linear function of the topographic index. Desbarats et al. (2002) applied KED to the interpolation of water table elevation using two deterministic trend models that include: (1) the topographic elevation and (2) the topographic index. Rivest et al. (2008) approximated the external drift using numerical solutions for the hydraulic head field obtained by means of finite elements based on a conceptual model that included hydrogeological parameter estimates, geology, and boundary conditions. KED and collocated CoK incorporated topography as secondary information in Boezio et al. (2006a,b). Both methodologies improved the quality of the water table elevation maps compared to OK. Another approach combines KED with the regionalized autoregressive exogenous variable model with precipitation surplus as the exogenous variable, and with DEM data as secondary variables (Knotters and Bierkens, 2002).

Neuman and Jacobsen (1984) used RK to estimate the hydraulic head in a catchment by approximating the trend function with space polynomials. RK with rainfall data as secondary variable was also applied to examine the influence of land use/cover change on the temporal and spatial variability of groundwater levels (Moukana and Koike, 2008). Nikroo et al. (2009) predicted water table elevation by different (SK, OK, and RK) kriging methods and trend functions, including auxiliary information from ground surface elevation and slope as well as draining rates.

Other research included kriging interpolation techniques in extended comparison studies regarding different interpolation methods applied to groundwater level data along with other hydrological variables, e.g., Subyani and Sen (1989), Kholghi and Hosseini (2009), and Sun et al. (2009).

It can be proved mathematically that KED and RK are practically equivalent, although the methodological steps differ (Hengl et al. 2003, 2007). The KED estimator is analyzed into a generalized regression of the primary variable with the secondary variables followed by SK or OK of the regression detrended residuals; in the kriging equation system the covariance matrix is extended with the auxiliary predictors. A limitation of KED is the potential instability of the extended matrix if the covariate varies irregularly in space (Goovaerts, 1997). In RK the drift model coefficients are first determined by regression, and the residuals are then interpolated using OK and finally added to the drift model. The main advantage of RK over KED is that it explicitly separates the trend estimation from the interpolation of the residuals, thus enabling the use of advanced regression methods (Hengl et al., 2003; Hengl, 2007). In addition, RK permits separate interpretation of the interpolated components and straightforward inclusion of multiple sources of external information that may compensate for small sample sizes (Alsamamra et al., 2009).

4. Overview of Geostatistical Methodology

In the following we will assume that the hydraulic head is represented by an SRF, which herein will generally be denoted by $Z(\mathbf{s},\omega)$, where ω is the state index used to denote that $Z(\mathbf{s},\omega)$ is a realization from an ensemble of possible states (to be omitted for brevity). The sampled field at the measurement points will be denoted by $Z(\mathbf{s}\in S)$, where S is the set of sampling points with cardinal number N. The values of the SRF in a given state will be denoted by lower-case letters. The target is to derive estimates, $\widehat{Z}(\mathbf{s}\in P)$, of the head at the prediction set points, P, that lie on a rectangular grid that covers the basin. Therefore $\mathbf{s}_i, i = 1,\ldots,N$ denote the sampling points, $z(\mathbf{s}_i)$ are the head values (in masl) at these points, and $\mathbf{s}_0$ denotes an estimation point, which is assumed to lie inside the convex hull of the sampling network. For mapping purposes, it is assumed that $\mathbf{s}_0$ moves sequentially through all the nodes of the mapping grid.

We examine linear interpolation methods for mapping spatial and/or temporal groundwater level variability. In spatial linear interpolation methods, it holds that:

$$\widehat{z}(\mathbf{s}_0) = \sum_{\{i\mathbf{s}_i \in \mathbb{S}_0\}} \lambda_i z(\mathbf{s}_i), \tag{1.27}$$

where $\mathbb{S}_0$ is the set of sampling points in the search neighborhood of $\mathbf{s}_0$. The neighborhood is empirically chosen so as to optimize the cross-validation measures.

For spatial interpolation we initially use two deterministic (IDW, MC) and three stochastic (OK, UK, DK) methods (Chapter 4). Then, we use OK (Chapter 5) and RK (Chapter 6) methods in combination with nonlinear normalizing transformations. In the first approach we apply a normalizing transformation $g(\cdot)$ to the data. Then we use OK to predict the transformed field $Y(\mathbf{s}) = g(Z(\mathbf{s}))$, and we back-transform the predictions to obtain head estimates. Several methods can be used to handle non-Gaussianity in the data. We applied the Box–Cox (BC) transformation, trans-Gaussian kriging (TGK), Gaussian anamorphosis (GA), and a modified Box–Cox (MBC) transform. We review these methods in the following chapters.

Herein we opt to keep the interpolation estimates within the convex hull of the sampling points. In principle we can estimate maps over the entire study domain; however, this is equivalent to extrapolation. Kriging can be used for extrapolation but the results outside the quadrilateral, determined from the sampling location boundaries, are often less accurate and subject to higher uncertainty. In addition, the semivariogram is determined by the measurements and expresses the spatial dependence of the measured points. In performing extrapolation, we accept that the semivariogram is valid outside the range of measurements. Therefore the estimates inside the quadrilateral are more accurate and precise than those outside.

5. Interpolation Materials and Methods

Interpolation is the process of estimating the data values in unvisited locations using known measured data values from neighbor points. The interpolation methods are divided between deterministic and stochastic. Deterministic methods provide no information regarding the possible estimation errors, while stochastic methods provide probabilistic estimates (i.e., provide the variance of the estimates). Deterministic interpolation methods assign weights to each sample value depending only on the distance between the sample point and the location of the interpolated point. On the other hand, stochastic or geostatistical methods treat the observations dataset as an arbitrary realization of a stochastic process and employ the spatial correlations between the values at neighboring points to distribute the weights. In this section the theoretical background of the deterministic and stochastic interpolation methods is explicitly presented (Varouchakis and Hristopulos, 2013).

5.1 Inverse Distance Weight

Estimation with the IDW method is given by means of the equation:

$$\hat{z}(\mathbf{s}_0) = \sum_{\{i\mathbf{s}_i \in \mathbb{S}_0\}} \left(\frac{d_{i,0}^{-n}}{\sum_{\{i\mathbf{s}_i \in \mathbb{S}_0\}} d_{i,0}^{-n}} \right) z(\mathbf{s}_i), \tag{1.28}$$

where $d_{i,0}$ is the distance between the estimation point and the sampling points and $n > 0$ is the power exponent; usually $n = 2$ is used. IDW assigns larger weights to data closer to the estimation point $\mathbf{s}_0$ than to more distant points. Higher values of n increase the impact of values near the interpolated point, while lower values of n imply more uniform weights. As it follows from Eq. (1.28) the weights add up to one. IDW is an exact and convex interpolation method (Hengl et al., 2007). In addition, it is very fast, straightforward, and computationally nonintensive (Webster and Oliver, 2001). According to Eq. (1.28), as the distance of $\mathbf{s}_i$ from $\mathbf{s}_0$ increases, the respective weight is reduced. IDW's disadvantages are the arbitrary choice of the weighting function and the lack of an uncertainty measure (Webster and Oliver, 2001).

5.2 Minimum Curvature

MC interpolation is based on the minimization of the total square curvature of the surface $z(\mathbf{s})$, i.e., $\int d\mathbf{s}\left[\nabla^2 z(\mathbf{s})\right]^2$, subject to the data constraints. In MC, the interpolated surface can be viewed as a thin linear elastic plate pinned to the data values at the sampling points. The estimate is obtained by solving the biharmonic partial differential equation (Briggs, 1974; Sandwell, 1987), i.e.:

$$\left(\frac{\partial^2}{\partial x^2} + \frac{\partial^2}{\partial y^2}\right)\left(\frac{\partial^2 z(\mathbf{s})}{\partial x^2} + \frac{\partial^2 z(\mathbf{s})}{\partial y^2}\right) = 0, \tag{1.29}$$

conditioned by the data values $z(\mathbf{s}_i)$. The interpolating function $z(\mathbf{s})$ honors the observed data and tends to a planar surface as the distance between the interpolation point and the observations increases. Typical applications of MC include interpolating hydrocarbon (oil) depths (Cooke et al., 1993), interpolation of gravitometric and magnetometric geophysical data for mineral exploration (Mendonca and Silva, 1995; Kay and Dimitrakopoulos, 2000), and mapping the earth's surface (Yilmaz, 2007).

The MC method often suffers from oscillations due to the presence of outliers in the data or due to very large gradients. This problem can become important if the dataset is relatively small. MC interpolation is based on Green's function g_m of the biharmonic equation, which satisfies $\nabla^4 g_m(\mathbf{s} - \mathbf{s}') = \delta(\mathbf{s} - \mathbf{s}')$ where $\delta(\mathbf{s} - \mathbf{s}')$ is the Dirac delta function. The two-dimensional Green's function is given by $g_m(d) = d^2(\ln d - 1)$ (Sandwell, 1987; Wessel, 2009). The MC estimate is then expressed as follows:

$$\hat{z}(\mathbf{s}_0) = \sum_{i=1}^{N} w_i g_m(d_{i,0}) \tag{1.30}$$

The weights w_i are determined by solving the following linear system at the N data locations:

$$z(\mathbf{s}_i) = \sum_{j=1}^{N} w_j g_m(d_{i,j}), \tag{1.31}$$

where $j = 1,\ldots,N$ and $d_{i,j}$ are the distances between the sample points $d_{i,j} = |\mathbf{s}_i - \mathbf{s}_j|$.

5.3 Ordinary Kriging Interpolation

The term kriging is used for a suite of interpolation methods that are based on the principles of zero bias and minimum mean square error. Kriging estimates the value of a process over an entire domain, over a finite-volume block, or at a specific point $\mathbf{s}_0$. The estimates are formed by means of a linear combination of the data values. Summation is over the entire area or a restricted neighborhood centered at the estimation point. The kriging interpolation method also quantifies the estimation variance, and thus the precision of the resulting estimates. The commonly used OK method is based on the following equations (Goovaerts, 1997; Kitanidis, 1997).

The OK method assumes that $z(\mathbf{s})$ is a random function with a constant but unknown mean. The OK estimate $\widehat{z}(\mathbf{s}_0)$ at $\mathbf{s}_0$ is calculated based on a weighted sum of the data:

$$\widehat{z}(\mathbf{s}_0) = \sum_{\{i\mathbf{s}_i \in \mathbb{S}_0\}} \lambda_i z_i(\mathbf{s}_i) \tag{1.32}$$

The weights λ_i in Eq. (1.32) are obtained by minimizing the mean square estimation error conditionally on the zero-bias constraint (Cressie, 1993), and they depend on the semivariogram model $\gamma_z(\mathbf{r})$ (Deutsch and Journel, 1992).

The kriging weights λ_i follow from the minimization of the mean square error and are given by the following $(N_0 + 1) \times (N_0 + 1)$ linear system of equations:

$$\sum_{\{i\mathbf{s}_i \in \mathbb{S}_0\}} \lambda_i \gamma_z(\mathbf{s}_i, \mathbf{s}_j) + \mu = \gamma_z(\mathbf{s}_j, \mathbf{s}_0), \quad j = 1, \ldots, N_0. \tag{1.33}$$

$$\sum_{\{i\mathbf{s}_i \in \mathbb{S}_0\}} \lambda_i = 1, \tag{1.34}$$

where N_0 is the number of points within the search neighborhood of $\mathbf{s}_0$, $\gamma_z(\mathbf{s}_i,\mathbf{s}_j)$ is the semivariogram between two sampled points $\mathbf{s}_i$ and $\mathbf{s}_j$, $\gamma_z(\mathbf{s}_j,\mathbf{s}_0)$ is the semivariogram between $\mathbf{s}_j$ and the estimation point $\mathbf{s}_0$, and μ is the Lagrange multiplier enforcing the no-bias constraint. $N_0 + 1, j = j, N_0 + 1 = 1$ for $j = 1,\ldots,N_0$, while $N_0 + 1, N_0 + 1 = 0$. Eq. (1.34) enforces the zero-bias condition.

Kriging provides not only an estimation of the variable $z(\mathbf{s}_0)$ but also the corresponding estimation's error variance (associated uncertainty). For OK the error variance (1) depends on the semivariogram model; the estimation precision depends on the complexity of the spatial variability of random field Z as modeled by the semivariogram, (2) depends on the data configurations and their distances to the location $z(\mathbf{s}_0)$ being estimated, (3) is independent of data values; for a given semivariogram model, two identical data configurations yield the same variance no matter their values, and (4) the error variance is zero at data locations and increases away from the data, while it reaches a maximum value for the extrapolation situation.

The OK estimation variance is defined by:

$$\sigma_E^2(\mathbf{s}_0) = E\left[\left\{Z(\mathbf{s}_0) - \widehat{Z}(\mathbf{s}_0)\right\}^2\right]$$

and is given by the following equation, with the Lagrange coefficient μ compensating for the uncertainty of the mean value:

$$\sigma_E^2(\mathbf{s}_0) = \sum_{is_i \in S_0} \lambda_i \gamma_z(\mathbf{s}_i, \mathbf{s}_0) + \mu. \tag{1.35}$$

Overall OK variance is termed as the weighted average of semivariograms from the new point $\mathbf{s}_0$ to all calibration points $\mathbf{s}_j$, plus the Lagrange multiplier.

5.4 R Script: Ordinary Kriging

The R programming environment has been used in this work to provide potential users with a free script where they can perform a complete geostatistical analysis of their datasets in terms of OK. The presented script has been formed, modified, assembled, and enriched appropriately using as basis the geoR package (Ribeiro, P.J., Diggle, P.J., 2001. geoR: a package for geostatistical analysis. *R News* 1(2), 15–18. http://CRAN.R-project.org/doc/Rnews/):

```
library(geoR)
library(fields)
library(maps)
#put ''yourdata.csv'' in the right path
dat <- read.csv("C:\\R\\yourdata.csv",header = TRUE)
x <- dat$x
y <- dat$y
z <- dat$z
s <- cbind(x,y)
#plot variogram
data_cloud <- variog(data=z,coords=s, option="cloud", max.dist=10000)
plot(data_cloud, scaled=TRUE, pch=16)
data_cloud_b <- variog(data=z,coords=s, bin.cloud=TRUE, max.dist=3000)
plot(data_cloud_b , bin.cloud=TRUE)
bins<- seq(0,3000,400)
vg <- variog(data=z,coords=s, uvec=bins)
plot(vg)
variofit1 <- variofit(vario=vg,cov.model="exponential",ini.cov.pars=c(50,1000),fix.
nugget=FALSE, nugget=0.75,weight="equal")
# for other covariance model types use "matern", "gaussian", "spherical", "circular",
"cubic", "wave", "power", "powered.exponential"
# draw lines on the empirical variogram
lines.variomodel(x=bins,cov.model="exp",
cov.pars=variofit1$cov.pars,nugget=variofit1$nugget,lty=2)
#cross validation
xv.cv <- xvalid(data=z,coords=s, model = variofit1)
plot(xv.cv,coords=s)
#Create grid of prediction points:
spl <- seq(min(x),max(x),length=100)
```

```
sp2 <- seq(min(y),max(y),length=100)
sp <- expand.grid(sp1,sp2)
sp
#Perform ordinary kriging interpolation:
pred<- krige.conv(data=z,coords=s,locations=sp, krige=krige.
control(cov.model="exponential", cov.pars=variofit1$cov.pars,
 nugget=variofit1$nugget,kappa=variofit1$kappa))
#Plot the ordinary kriging predicted values:
image.plot(sp1,sp2,matrix(pred$predict,100,100),main="Prediction Values")
map("county",add=T)
ch <- chull(s)
coords <- dat[c(ch, ch[1]), ] # closed polygon
a=coords[,c(1)]
b=coords[,c(2)]
aa<- cbind(a,b)
bor<- aa
kc=pred
image(kc, borders=bor)
points(s,pch=19)
contour(kc, borders=bor, add=TRUE)
# Plot the ordinary kriging standard deviation:
image.plot(sp1,sp2,matrix(sqrt(pred$krige.var),100,100),main="Standard Deviation")
map("county",add=T)
ch <- chull(s)
coords <- dat[c(ch, ch[1]), ] # closed polygon
a=coords[,c(1)]
b=coords[,c(2)]
aa<- cbind(a,b)
bor<- aa
kc=pred
image(kc, val=sqrt(kc$krige.var), main="kriging std. errors", borders=bor)
points(s,pch=19)
```

5.5 Universal Kriging Interpolation

In certain cases, the data exhibit a global trend over the study area. It is possible to incorporate in kriging a trend (drift function) modeling the global behavior. The resulting estimation algorithm is known as UK and was proposed by Matheron (1969). UK requires the drift function $m_z(\mathbf{s})$ and the semivariogram of the residuals $e_z(\mathbf{s})$ (Goovaerts, 1997). The trend is usually approximated by linear or higher-order polynomials of the space coordinates (Ahmed, 2007). The drift function is given by:

$$m_z(\mathbf{s}) = \sum_{k=1}^{K} a_k f_k(\mathbf{s}), \tag{1.36}$$

where $f_k(\mathbf{s})$ are basis functions and a_k are the drift coefficients (Goovaerts, 1997). The UK estimator of the hydraulic head is expressed as follows:

$$\hat{z}(\mathbf{s}_0) = m_z(\mathbf{s}_0) + \sum_{\{i\mathbf{s}_i \in \mathbb{S}_0\}} \lambda_i e(\mathbf{s}_i) = m_z(\mathbf{s}_0) + \sum_{\{i\mathbf{s}_i \in \mathbb{S}_0\}} \lambda_i [z(\mathbf{s}_i) - m_z(\mathbf{s}_i)], \tag{1.37}$$

where λ_i $(i = 1, \ldots, N_0)$ are the UK weights, $e(\mathbf{s}_i)$is the residual at $\mathbf{s}_i$, and $m_z(\mathbf{s}_0)$ is the drift at $\mathbf{s}_0$.

The kriging weights are determined by the solution of the following $(N_0 + K) \times (N_0 + K)$ linear system of equations, where N_0 is the number of points within the search neighborhood of $\mathbf{s}_0$:

$$\sum_{\{i\mathbf{s}_i \in \mathbb{S}_0\}} \lambda_i \gamma_z(\mathbf{s}_i, \mathbf{s}_j) + \sum_{k=1}^{K} f_k(\mathbf{s}_j)\mu_k = \gamma_z(\mathbf{s}_j, \mathbf{s}_0), \quad j = 1, \ldots, N_0 \tag{1.38}$$

$$\sum_{\{i\mathbf{s}_i \in \mathbb{S}_0\}} \lambda_i f_k(\mathbf{s}_i) = f_k(\mathbf{s}_0), \quad k = 1, \ldots, K, \tag{1.39}$$

where $\gamma_z(\mathbf{s}_i,\mathbf{s}_j)$ is the semivariogram of the residuals between two sampled points $\mathbf{s}_i$ and $\mathbf{s}_j$, $\gamma_z(\mathbf{s}_j,\mathbf{s}_0)$ is the semivariogram of the residuals between a sampled point $\mathbf{s}_j$ and the estimation point $\mathbf{s}_0$, and μ_k are the Lagrange multipliers for each basis function. The kriging variance is given by the following equation (Goovaerts, 1997):

$$\sigma_E^2(\mathbf{s}_0) = \sum_{\{i\mathbf{s}_i \in \mathbb{S}_0\}} \lambda_i \gamma_z(\mathbf{s}_i, \mathbf{s}_0) + \sum_{k=1}^{K} f_k(\mathbf{s}_0)\mu_k \tag{1.40}$$

5.6 Kriging With Delaunay Triangulation

DK uses the Delaunay triangles to determine the search neighborhood $\mathbb{S}_0$ around the estimation point. The kriging equations in DK are identical to OK (Hessami et al., 2001). DK reduces the computational cost of kriging and ensures that the estimate depends only on data in each point's immediate neighborhood.

The Delaunay triangulation (e.g., Fig. 1.4) is the dual graph of the Voronoi diagram for the sampling locations $\mathbf{s}_i$, $i = 1,\ldots,N$. The latter is a set of polygons P_i, each of which is centered at $\mathbf{s}_i$ and contains all the points that are closer to $\mathbf{s}_i$ than to any other data point. The Delaunay triangulation is formed by drawing line segments between Voronoi vertices if their respective polygons have a common edge (Okabe et al., 1992; Mulchrone, 2003; Ling et al., 2005). The Delaunay triangle containing the estimation point $\mathbf{s}_0$ is located using the T-search[1] function of Matlab (Matlab v.7.5). The vertices of the triangle T_0 containing $\mathbf{s}_0$ are the first-order neighbors of $\mathbf{s}_0$. Second-order neighbors are determined from the vertices of the triangles adjacent to T_0 that do not belong to T_0 (Hessami et al., 2001; see Fig. 1.4). The number of second-order neighbors ranges between one and

[1]The tsearch function will be replaced in future Matlab releases by DelaunayTri class.

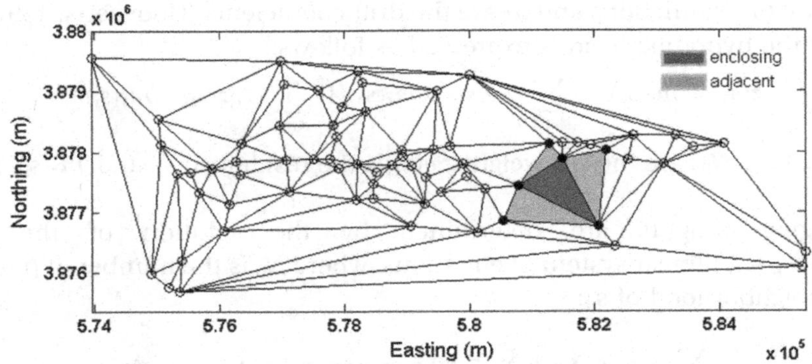

FIGURE 1.4 Example of Delaunay triangulation in monitoring sites. The vertices of the enclosing triangle (*dark color*) that contains the estimation point $\mathbf{s}_0$ are the first-order neighbors of $\mathbf{s}_0$; the vertices of the three adjacent triangles (*gray color*) that do not belong to the enclosing triangle provide the second-order neighbors of $\mathbf{s}_0$.

three. If the search neighborhood only includes the first-order neighbors, the CPU time is reduced but the precision of the estimates is lower (Hessami et al., 2001).

5.7 Residual Kriging

RK combines a trend function with interpolation of the residuals. In RK the estimate is expressed as:

$$\widehat{z}(\mathbf{s}_0) = m_z(\mathbf{s}_0) + \widehat{z}'(\mathbf{s}_0), \tag{1.41}$$

where $m_z(\mathbf{s}_0)$ is the trend function and $\widehat{z}'(\mathbf{s}_0)$ is the interpolated residual by means of OK (Rivoirard, 2002). Typically, the trend function is modeled as:

$$m_z(\mathbf{s}_0) = \sum_{k=0}^{p} \beta_k q_k(\mathbf{s}_0) q_k(\mathbf{s}_0) \equiv 1, \tag{1.42}$$

where $q_k(\mathbf{s}_0)$ are the values of *auxiliary variables* at $\mathbf{s}_0$, β_k are the estimated regression coefficients, and p is the number of auxiliary variables (Draper and Smith, 1981; Hengl, 2007; Hengl et al., 2007). Auxiliary variables could include polynomials of the data coordinates (x,y). The regression coefficients are estimated from the sample using ordinary least squares or generalized least squares (GLS). However, it has been shown (Kitanidis, 1993) that GLS does not confer any significant benefit if the sampling locations are not clustered. The variance of the estimates follows from the equations (Hengl et al. 2003, 2007):

$$\sigma^2(\mathbf{s}_0) = \sigma_d^2(\mathbf{s}_0) + \sigma_f^2(\mathbf{s}_0), \tag{1.43}$$

$$\sigma_d^2(\mathbf{s}_0) = \mathbf{q}_0^T \left(\mathbf{q}^T \gamma_{z'}^{-1} \mathbf{q}\right)^{-1} \mathbf{q}_0, \tag{1.44}$$

$$\sigma_f^2(\mathbf{s}_0) = \sum_{\{i\mathbf{s}_i \in \mathbb{S}_0\}} \lambda_i \gamma_{z'}(\mathbf{s}_i, \mathbf{s}_0) + \mu, \tag{1.45}$$

where σ_d^2 $(\mathbf{s}_0)$ is the drift prediction variance, $\mathbf{q}_0$ is the vector of $(p+1) \times 1$ predictors at the unvisited location, $\mathbf{q}$ is the matrix of $(N_0+1) \times (p+1)$ predictors at the sampling points in the search neighborhood, $\gamma_{z'}$ is the semivariogram matrix of the $(N_0+1) \times (N_0+1)$ residuals at the measured locations (neighborhood), and $\sigma_f^2(\mathbf{s}_0)$ is the kriging (OK) variance of residuals. The terms involved in the drift variance prediction are presented next in vector and matrix form as appropriate:

$$\mathbf{q}_0 = \begin{bmatrix} q_1(\mathbf{s}_0) \\ q_2(\mathbf{s}_0) \\ \vdots \\ q_p(\mathbf{s}_0) \\ 1 \end{bmatrix}$$

$$q = \begin{bmatrix} q_1(\mathbf{s}_1) & \cdots & \cdots & q_p(\mathbf{s}_1) & 1 \\ q_1(\mathbf{s}_2) & \cdots & \cdots & q_p(\mathbf{s}_2) & 1 \\ \vdots & \vdots & \vdots & \vdots & \vdots \\ q_1(\mathbf{s}_{N_0}) & \cdots & \cdots & q_p(\mathbf{s}_{N_0}) & 1 \\ 1 & \cdots & \cdots & 1 & 1 \end{bmatrix} \tag{1.46}$$

$$\gamma_{z'} = \begin{bmatrix} \gamma_{z'}(\mathbf{s}_1, \mathbf{s}_1) & \cdots & \cdots & \gamma_{z'}(\mathbf{s}_1, \mathbf{s}_{N_0}) & 1 \\ \gamma_{z'}(\mathbf{s}_2, \mathbf{s}_1) & \cdots & \cdots & \gamma_{z'}(\mathbf{s}_1, \mathbf{s}_2) & 1 \\ \vdots & \vdots & \vdots & \vdots & \vdots \\ \gamma_{z'}(\mathbf{s}_{N_0}, \mathbf{s}_1) & \cdots & \cdots & \gamma_{z'}(\mathbf{s}_{N_0}, \mathbf{s}_{N_0}) & 1 \\ 1 & \cdots & \cdots & 1 & 0 \end{bmatrix}.$$

6. Spatial Model Validation

Leave-one-out cross-validation is usually applied to compare the different spatial models. This procedure consists of removing one datum at a time from S and estimating its value based on the remaining $N-1$ data. Interpolated values are compared with their measured counterparts using the global performance measures listed next. The "optimal" spatial model is determined based on the performance of statistical metrics that quantify differences between the estimated and true values. The validation measures defined next are used, where $z^*(\mathbf{s}_i)$ and $z(\mathbf{s}_i)$ are, respectively, the estimated and true head values at point $\mathbf{s}_i$. The estimates are obtained by removing $z(\mathbf{s}_i)$ from the dataset and interpolating the remaining data; $\overline{z(\mathbf{s}_i)}$ denotes the spatial average of the data and $\overline{z^*(\mathbf{s}_i)}$ the spatial average of the estimates, while N is the number of observations.

Mean absolute error:

$$\varepsilon_{\mathrm{MA}} = \frac{1}{N}\sum_{i=0}^{N}\left|z^*(\mathbf{s}_i) - z(\mathbf{s}_i)\right|, \tag{1.47}$$

Bias:

$$\varepsilon_{\mathrm{BIAS}} = \frac{1}{N}\sum_{i=0}^{N} z^*(\mathbf{s}_i) - z(\mathbf{s}_i), \tag{1.48}$$

Mean absolute relative error:

$$\varepsilon_{\mathrm{MAR}} = \frac{1}{N}\sum_{i=0}^{N}\left|\frac{z^*(\mathbf{s}_i) - z(\mathbf{s}_i)}{z(\mathbf{s}_i)}\right|, \tag{1.49}$$

Root mean square error:

$$\varepsilon_{\mathrm{RMS}} = \sqrt{\frac{1}{N}\sum_{i=0}^{N}\left[z^*(\mathbf{s}_i) - z(\mathbf{s}_i)\right]^2}, \tag{1.50}$$

Linear correlation coefficient:

$$R = \frac{\sum_{i=0}^{N}\left[z(\mathbf{s}_i) - \overline{z(\mathbf{s}_i)}\right]\left[z^*(\mathbf{s}_i) - \overline{z^*(\mathbf{s}_i)}\right]}{\sqrt{\sum_{i=1}^{N}\left[z(\mathbf{s}_i) - \overline{z(\mathbf{s}_i)}\right]^2}\sqrt{\sum_{i=1}^{N}\left[z^*(\mathbf{s}_i) - \overline{z^*(\mathbf{s}_i)}\right]^2}}, \tag{1.51}$$

7. Comparison of Interpolation Methods

There is no universally optimum interpolation method that can be used for all kinds of datasets. Two spatial interpolation comparison exercises have been organized by the Radioactivity Environmental Monitoring Group of the Joint Research Centre of the European Commission (Dubois, 1998; Dubois and Galmarini, 2005). These exercises focused on radioactivity monitoring in the European continent and in particular on automatic (i.e., without user involvement) mapping.

Cornford (2005) emphasized the problems of interpreting comparative interpolation studies. First, their results do not admit generalization and are often contradictory. In addition, for a single dataset several or all assessed methods may exhibit similar performance. Hence the choice of the "optimal" interpolation method is dictated by other factors, such as computational speed, implementation cost, scaling with data size, and the ability to make probabilistic predictions (estimates of the prediction error). Van den Boogaart (2005) agrees that comparative studies based on one or two datasets can be misleading, and that a uniformly optimal method for all kinds of dataset does not exist. He points out that the performance and utility of the methods should be assessed in terms of decision-making requirements (e.g., concerning outliers, estimation variances) and its adaptability to the complexity of the specific dataset (e.g., sparse data, presence of trends) and not only in terms of mean square errors. Myers (2005) emphasizes the use of clear software standards, common hardware configurations, and an extensive set of performance measures to allow the duplication of reported results by others. In light of the foregoing remarks, the same programming environment has to be applied for all the methods tested so that results are directly comparable.

8. Improving Kriging Using Nonlinear Normalizing Transformation

Skewed or erratic data can often be made more suitable for geostatistical modeling by appropriate transformation. Such applications can lead to reliable mapping of environmental data. A normal distribution for the variable under study is desirable in linear geostatistics (Clark and Harper, 2000). Even though mild deviations from normality do not cause problems, significant deviations, e.g., due to high skewness and outliers, have an undesirable impact on the semivariogram structure and the kriging estimates (Gringarten and Deutsch, 2001; Ouyang et al., 2006). OK is well known to be optimal when the data have a multivariate normal distribution and the true semivariogram is known. Therefore transformation of data may be required before kriging to normalize the data distribution, suppress outliers, and improve data stationarity (Deutsch and Journel, 1992; Armstrong, 1998). Then the estimation is performed in the Gaussian domain before back-transforming the estimates to the original domain. An advantage of the Gaussian distribution is that spatial variability is easier to model, because it reduces effects of extreme values providing more stable semivariograms (Goovaerts, 1997; Armstrong, 1998; Pardo-Iguzquiza and Dowd, 2005). Kriging represents variability only up to the second-order moment (covariance), therefore the random field of the transformed variable must be Gaussian to derive unbiased estimates at nonsampled locations (Deutsch and Journel, 1992; Goovaerts et al., 2005). In practice, multinormality is invoked as a working hypothesis.

The aim of this work is to investigate the improvement of interpolation with OK using nonlinear data normalization methodologies. Well-known OK-based methodologies are applied such as the MBC method, the GA normalization method, and the TGK method. In addition, a novel normalization method based on BC transformation, termed MBC, is herein established and implemented (Varouchakis et al., 2012).

9. Box–Cox Transformation Method

The BC method (Box and Cox, 1964) is widely used to transform hydrological data into approximately Gaussian distributions (Chander et al., 1978; Hirsch, 1979; Jain and Singh,

1986; Salas, 1993; Thyer et al., 2002). The transform is defined only for positive data values and is defined by means of:

$$y := g_{\mathrm{BC}}(z;k) = \begin{cases} \dfrac{(z^k - 1)}{k}, & k \neq 0 \\ \log(z), & k = 0 \end{cases} \tag{1.52}$$

Given the vector of data observations $\mathbf{z}^T = (z_1, \ldots, z_N)$, the optimal value of the power exponent k, which leads to the best agreement of $\mathbf{y}^T = (g_k(z_1), \ldots, g_k(z_N))$ with the Gaussian distribution, can be determined by means of the maximum likelihood estimation method (De Oliveira et al., 1997). The power exponent k is estimated by maximizing the logarithm of the likelihood function:

$$f(\mathbf{z};k) = -\frac{N}{2}\ln\left[\sum_{i=1}^{N} \frac{\left(g_{\mathrm{BC}}(z_i;k) - \overline{g}_{\mathrm{BC}}(\mathbf{z};k)\right)^2}{N}\right] + (k-1)\sum_{i=1}^{N} \ln(z_i), \tag{1.53}$$

where $\overline{g}_{\mathrm{BC}}(\mathbf{z};k) = \frac{1}{N}\left[\sum_{i=1}^{N} g_{\mathrm{BC}}(z_i;k)\right]$ is the arithmetic mean of the transformed data, while the sum of squares $\left[\sum_{i=1}^{N} \frac{\left(g_{\mathrm{BC}}(z_i;k) - \overline{g}_{\mathrm{BC}}(z_i;k)\right)^2}{N}\right]$ denotes the transformed data variance.

10. Trans-Gaussian Kriging

TGK is more general than the BC transformation (Cressie, 1993; Kozintseva, 1999; Schabenberger and Gotway, 2005). For a nonlinear normalizing transformation, $Y(\mathbf{s}) = g(Z(\mathbf{s}))$, where $Y(\mathbf{s})$ follows the multivariate Gaussian distribution, assuming that $Z(\mathbf{s}) = \varphi(Y(\mathbf{s}))$, where $\varphi(\bullet) = g^{-1}(\bullet)$ is a one-to-one, twice-differentiable function. It is also assumed that $Y(\mathbf{s})$ is an intrinsically stationary SRF with mean m_Y and semivariogram $\gamma_Y(\mathrm{r})$. For an unknown m_Y, the OK predictor, $\widehat{Y}_{\mathrm{OK}}(\mathbf{s}_0)$, is used to predict $Y(\mathbf{s}_0)$. An estimate of $Z(\mathbf{s}_0)$ is then given by $\widehat{Z}(\mathbf{s}_0) = \varphi\left(\widehat{Y}_{\mathrm{OK}}(\mathbf{s}_0)\right)$, where $\varphi(\bullet)$is the inverse of the transformation function. However, this results in a biased predictor if $\varphi(\bullet)$ is a nonlinear transformation. A bias-correcting approximation is the trans-Gaussian predictor (Cressie, 1993):

$$\widehat{Z}(\mathbf{s}_0) = \varphi\left(\widehat{Y}_{\mathrm{OK}}(\mathbf{s}_0)\right) + \frac{\varphi''(\widehat{m}_Y)}{2}\left[\sigma^2_{\mathrm{OK};Y}(\mathbf{s}_0) - 2\mu_Y\right], \tag{1.54}$$

where $\hat{m}_Y$ is the OK estimate of m_Y, μ_Y is the Lagrange multiplier of the OK system, $\varphi''(\cdot)$ is the second-order derivative of the inverse transformation function, and $\sigma^2_{\mathrm{OK};Y}(\mathbf{s}_0)$ is the OK variance. If the BC normalizing transformation Eq. (1.52) is used, as herein, the functions $\varphi(\cdot)$ and $\varphi''(\cdot)$ have the following form:

$$\varphi(y) = (ky+1)^{1/k}, \tag{1.55}$$

$$\varphi''(\widehat{m}_Y) = (1-k)(\widehat{m}_Y k + 1)^{\frac{1}{k}-2}. \tag{1.56}$$

11. Gaussian Anamorphosis

This method is based on the transformation of a Gaussian variable Y into a new variable Z with an arbitrary distribution by means of $Z = \Phi_{GA}(Y)$, where $\Phi_{GA}(\cdot)$ is the Gaussian anamorphosis transformation. The transformation used in GA involves the following polynomial expansion (Chiles and Delfiner, 1999):

$$\Phi_{GA}(Y) = \sum_{i=0}^{K} \Psi_i H_i(Y), \tag{1.57}$$

where the functions $H_i(Y)$, $i = 0,\ldots,K$ are *Hermite polynomials* and Ψ_i denotes the coefficients of the expansion. The Hermite polynomials are defined in terms of the derivatives of the Gaussian density function:

$$H_i(x) = \frac{g^{(i)}(x)}{g(x)}, \tag{1.58}$$

where $g(x)$ is the zero-mean and unit variance Gaussian density function, i.e., $g(x) = \frac{1}{\sqrt{2\pi}} e^{-\frac{x^2}{2}}$, and $g^{(i)}(x)$ is the ith-order derivative of $g(x)$. The Hermite polynomials are calculated by means of the following recurrence relation:

$$H_{i+1}(x) = -xH_i(x) - iH_{i-1}(x), \quad i \geq 0. \tag{1.59}$$

Typically, a high polynomial order ($K = 30-100$) is used. Model fitting consists of estimating the coefficients Ψ_i. The normalization of a non-Gaussian variable requires the inversion of the anamorphosis function by means of $Y = \Phi_{GA}^{-1}(Z)$. The geostatistical analysis is performed on the transformed variable Y, and the estimates are finally back-transformed to the original values through the anamorphosis function (Olea, 1999; Wackernagel, 2003; Casa and Castrignano, 2008).

Practically any function of Y that is square integrable with respect to the Gaussian density can be expanded in terms of Hermite polynomials. The coefficients of the expansion are given by the following equation (Journel and Huijbregts, 1978; Wackernagel, 2003):

$$\Psi_i = \int_{-\infty}^{\infty} \Phi_{GA}(x) H_i(x) g(x) dx. \tag{1.60}$$

The expansion coefficients Ψ_i are estimated for the linear, polynomial, and exponential functions. The function Φ_{GA} that gives the best fit to the data is the quadratic function $\Phi_{GA}(x) = x^2$. For the quadratic, the integral Eq. (1.60) used to estimate Ψ_i is solved analytically for any Hermite polynomial using integration by parts. As an example, for the second-order Hermite polynomial, Eq. (1.60) becomes:

$$\Psi_2 = \frac{1}{\sqrt{2\pi}} \left(\int_{-\infty}^{\infty} x^4 e^{-\frac{x^2}{2}} dx - \int_{-\infty}^{\infty} x^2 e^{-\frac{x^2}{2}} dx \right) = 2. \tag{1.61}$$

In general, the solution of the integral is:

$$a_n = \int_{-\infty}^{\infty} x^n e^{-\frac{x^2}{2}} dx = \sqrt{2\pi}(n-1)!! = \sqrt{2\pi}(n-1)(n-3)\cdots 3 \times 1, \tag{1.62}$$

for n even, while $a_n = 0$ for odd n. Hence the corresponding expansion coefficients Ψ_i vanish for Hermite polynomials of odd order.

12. Modified Box–Cox

This new method focuses on normalizing the skewness and kurtosis coefficients of the data, but it neglects higher-order moments. It is defined by the following function:

$$y := g_{\text{MBC}}(z; \boldsymbol{\kappa}) = \frac{\left(z - z_{\min} + k_2^2\right)^{k_1} - 1}{k_1}, \boldsymbol{\kappa}^T = (k_1, k_2), \tag{1.63}$$

where k_1 is the power exponent and k_2 is an offset parameter. Use of the latter allows negative z values and so the transformation in Eq. (1.63) can be applied to fluctuations as well. Parameters (k_1,k_2) are estimated from the numerical solution of the equations $\widehat{s}_z = 0, \widehat{k}_z = 3$, where $\widehat{s}_z$ and $\widehat{k}_z$ are the sample skewness and kurtosis coefficients, respectively:

$$\left(\frac{\widehat{m}_z - \widetilde{m}_z}{\sigma_z}\right)^2 + \left(\widehat{k}_z - 3\right)^2 \simeq 0, \tag{1.64}$$

where $\widetilde{m}_z$ is the sample's median. Minimization is performed using the Nelder–Mead simplex optimization method (Nelder and Mead, 1965; Press et al., 1992).

References

Abedini, M.J., Nasseri, M., Ansari, A., 2008. Cluster-based ordinary kriging of piezometric head in West Texas/New Mexico—testing of hypothesis. Journal of Hydrology 351 (3–4), 360–367.

Aboufirassi, M., Marino, M., 1983. Kriging of water levels in the Souss aquifer, Morocco. Mathematical Geology 15 (4), 537–551.

Ahmadi, S., Sedghamiz, A., 2007. Geostatistical analysis of spatial and temporal variations of groundwater level. Environmental Monitoring and Assessment 129 (1), 277–294.

Ahmadi, S., Sedghamiz, A., 2008. Application and evaluation of kriging and cokriging methods on groundwater depth mapping. Environmental Monitoring and Assessment 138 (1), 357–368.

Ahmed, S., 2007. Application of geostatistics in hydrosciences. In: Thangarajan, M. (Ed.), Groundwater. Springer, Netherlands, pp. 78–111.

Ahmed, S., De Marsily, G., 1987. Comparison of geostatistical methods for estimating transmissivity using data on transmissivity and specific capacity. Water Resources Research 23 (9), 1717–1737.

Alsamamra, H., Ruiz-Arias, J.A., Pozo-Vazquez, D., Tovar-Pescador, J., 2009. A comparative study of ordinary and residual kriging techniques for mapping global solar radiation over southern Spain. Agricultural and Forest Meteorology 149 (8), 1343–1357.

Armstrong, M., 1998. Basic Linear Geostatistics. Springer Verlag, Berlin.

Atkinson, P.M., Lloyd, C.D., 2010. Geostatistics for Environmental Applications. Springer Verlag.

Beven, K.J., Kirkby, M.J., 1979. A physically based, variable contributing area model of basin hydrology. Hydrological Sciences Bulletin 24 (1), 43–69.

Bochner, S., 1959. Lectures on Fourier Integrals. Princeton University Press, Princeton, NJ.

Boezio, M., Costa, J., Koppe, J., 2006a. Accounting for extensive secondary information to improve watertable mapping. Natural Resources Research 15 (1), 33–48.

Boezio, M., Costa, J., Koppe, J., 2006b. Kriging with an external drift versus collocated cokriging for water table mapping. Applied Earth Science 115 (3), 103–112.

Box, G.E.P., Cox, D.R., 1964. An analysis of transformations. Journal of the Royal statistical Society, Ser. B 26 (2), 211–252.

Briggs, I.C., 1974. Machine contouring using minimum curvature. Geophysics 39 (1), 39–48.

Buchanan, S., Triantafilis, J., 2009. Mapping water table depth using geophysical and environmental variables. Groundwater 47 (1), 80–96.

Burrough, P.A., 2001. GIS and geostatistics: essential partners for spatial analysis. Environmental and Ecological Statistics 8 (4), 361–377.

Casa, R., Castrignano, A., 2008. Analysis of spatial relationships between soil and crop variables in a durum wheat field using a multivariate geostatistical approach. European Journal of Agronomy 28 (3), 331–342.

Chander, S., Kumar, A., Spolia, S.K., 1978. Flood frequency analysis by power transformation. Journal of the Hydraulics Division: Proceedings ASCE 104 (11), 1495–1504.

Chiles, J.P., Delfiner, A., 1999. Geostatistics (Modeling Spatial Uncertainty). Wiley, New York.

Christakos, G., 1991. Random Field Models in Earth Sciences. Academic Press, San Diego.

Christakos, G., Hristopulos, D.T., 1998. Spatiotemporal Environmental Health Modelling: A Tractatus Stochasticus. Kluwer, Boston.

Clark, I., Harper, W.V., 2000. Practical Geostatistics 2000. Ecosse North America LLC, Columbus, Ohio, USA.

Cooke, R., Mostaghimi, S., Parker, J.C., 1993. Estimating oil spill characteristics from oil heads in scattered monitoring wells. Environmental Monitoring and Assessment 28 (1), 33–51.

Cornford, D., 2005. Are comparative studies a waste of time? SIC2004. In: Dubois, G. (Ed.), Automatic Mapping Algorithms for Routine and Emergency Monitoring Data. EUR 21595 EN – Scientific and Technical Research Series. Office for Official Publications of the European Communities, Luxembourg, ISBN 92-894-9400-X, pp. *61–70.* EUR, 2005 ed.

Cressie, N., 1993. Statistics for Spatial Data, revised ed. Wiley, New York.

Davis, J.C., 1973. Statistics and Data Analysis in Geology. Wiley, New York.

De-Vitry, C., Vann, J., Arvidson, H., 2010. Multivariate iron ore deposit resource estimation—a practitioner's guide to selecting methods. Transactions of the Institutions of Mining and Metallurgy, Section B: Applied Earth Science 119 (3), 154–165.

De Oliveira, V., Kedem, B., Short, D.A., 1997. Bayesian prediction of transformed gaussian random fields. Journal of the American Statistical Association 92 (440), 1422–1433.

Delhomme, J.P., 1974. La cartographie d'une grandeur physique a partir des donnees de differentes qualities. In: International Association of Hydrogeologists. Proc. of IAH Congress, Montpelier, France, pp. 185–194. Montpelier, France: IAH.

Delhomme, J.P., 1978. Kriging in the hydrosciences. Advances in Water Resources 1 (5), 251–266.

Desbarats, A.J., Logan, C.E., Hinton, M.J., Sharpe, D.R., 2002. On the kriging of water table elevations using collateral information from a digital elevation model. Journal of Hydrology 255 (1–4), 25–38.

Deutsch, C.V., Journel, A.G., 1992. GSLIB. Geostatistical Software Library and User's Guide. Oxford University Press, New York.

Draper, N., Smith, H., 1981. In: Applied Regression Analysis, second ed. Wiley, New York.

Dubois, G., 1998. Spatial interpolation comparison 97: foreword and introduction. Journal of Geographical Information and Decision Analysis 2 (2), 1–10.

Dubois, G., Galmarini, S., 2005. Spatial interpolation comparison SIC2004: introduction to the exercise and overview on the results. In: Dubois, G. (Ed.), Automatic Mapping Algorithms for Routine and Emergency Monitoring Data. EUR 21595 EN – Scientific and Technical Research Series. Office for Official Publications of the European Commission, Luxembourg, ISBN 92-894-9400-X, pp. *7–18*. EUR, 2005 ed.

Elogne, S., Hristopulos, D., Varouchakis, E., 2008. An application of Spartan spatial random fields in environmental mapping: focus on automatic mapping capabilities. Stochastic Environmental Research and Risk Assessment 22 (5), 633–646.

Gambolati, G., Volpi, G., 1979a. A conceptual deterministic analysis of the kriging technique in hydrology. Water Resources Research 15 (3), 625–629.

Gambolati, G., Volpi, G., 1979b. Groundwater contour mapping in Venice by stochastic interpolators 1. Theory. Water Resources Research 15 (2), 281–290.

Goovaerts, P., 1997. Geostatistics for Natural Resources Evaluation. Oxford University Press, New York.

Goovaerts, P., 1999. Using elevation to aid the geostatistical mapping of rainfall erosivity. Catena 34 (3–4), 227–242.

Goovaerts, P., AvRuskin, G., Meliker, J., Slotnick, M., Jacquez, G., Nriagu, J., 2005. Geostatistical modeling of the spatial variability of arsenic in groundwater of southeast Michigan. Water Resources Research 41, W07013. https://doi.org/10.1029/2004WR003705).

Gringarten, E., Deutsch, C.V., 2001. Teacher's aide: variogram interpretation and modeling. Mathematical Geology 33, 507–534.

Hengl, T., 2007. A Practical Guide to Geostatistical Mapping of Environmental Variables, vol. 143. Office for Official Publications of the European Communities, Luxembourg.

Hengl, T., Heuvelink, G.B.M., Rossiter, D.G., 2007. About regression-kriging: from equations to case studies. Computers & Geosciences 33 (10), 1301–1315.

Hengl, T., Heuvelink, G.B.M., Stein, A., 2003. Comparison of kriging with external drift and regression-kriging. International Institute for Geo-information Science and Earth Observation (ITC) 17.

Hessami, M., Anctil, F., Viau, A.A., 2001. Delaunay implementation to improve kriging computing efficiency. Computers & Geosciences 27 (2), 237–240.

Hirsch, R.M., 1979. Synthetic hydrology and water supply reliability. Water Resources Research 15 (6), 1603–1615.

Hoeksema, R.J., Clapp, R.B., Thomas, A.L., Hunley, A.E., Farrow, N.D., Dearstone, K.C., 1989. Cokriging model for estimation of water table elevation. Water Resources Research 25 (3), 429–438.

Hohn, M.E., 1999. Geostatistics and Petroleum Geology. Springer, Dordrecht, The Netherlands.

Hristopulos, D.T., 2002. New anisotropic covariance models and estimation of anisotropic parameters based on the covariance tensor identity. Stochastic Environmental Research and Risk Assessment 16 (1), 43–62.

Hristopulos, D.T., 2003a. Introduction to Geostatistics-Cource Notes (In Greek), vol. 200. Technical University of Crete, Chania, Crete, Greece.

Hristopulos, D.T., 2003b. Spartan Gibbs random field models for geostatistical applications. SIAM Journal on Scientific Computing 24 (6), 2125–2162.

Hristopulos, D.T., 2008. Applied Geostatistics-Cource Notes (In Greek), vol. 200. Technical University of Crete, Chania, Crete, Greece.

Hristopulos, D.T., Elogne, S.N., 2007. Analytic properties and covariance functions for a new class of generalized Gibbs random fields. IEEE Transactions on Information Theory 53 (12), 4667–4679.

Hristopulos, D.T., Elogne, S.N., 2009. Computationally efficient spatial interpolators based on spartan spatial random fields. IEEE Transactions on Signal Processing 57 (9), 3475–3487.

Isaaks, E.H., Srivastava, R.M., 1989. An Introduction to Applied Geostatisics. Oxford University Press, New York.

Jain, D., Singh, V.P., 1986. A comparison of transformation methods for flood frequency analysis. Water Resources Bulletin 22 (6), 903–912.

Journel, A.G., Huijbregts, C., 1978. Mining Geostatistics. Academic Press, New York.

Jowett, G.H., 1955. Sampling properties of local statistics in stationary stochastic series. Biometrika 42, 160–169.

Kay, M., Dimitrakopoulos, R., 2000. Integrated interpolation methods for geophysical data: applications to mineral exploration. Natural Resources Research 9 (1), 53–64.

Kholghi, M., Hosseini, S., 2009. Comparison of groundwater level estimation using neuro-fuzzy and ordinary kriging. Environmental Modeling & Assessment 14 (6), 729–737.

Kitanidis, P., 1993. Generalized covariance functions in estimation. Mathematical Geology 25 (5), 525–540.

Kitanidis, P.K., 1997. Introduction to Geostatistics. Cambridge University Press, Cambridge.

Knotters, M., Bierkens, M.F.P., 2002. Accuracy of spatio-temporal RARX model predictions of water table depths. Stochastic Environmental Research and Risk Assessment 16 (2), 112–126.

Knotters, M., Brus, D.J., Oude Voshaar, J.H., 1995. A comparison of kriging, co-kriging and kriging combined with regression for spatial interpolation of horizon depth with censored observations. Geoderma 67 (3–4), 227–246.

Kolmogorov, A.N., 1941. Interpolirovanie i ekstrapolirovanie statsionarnykh sluchainykh posledovatel' nostei (Interpolated and extrapolated stationary random sequences). Isvestia AN SSSR, Seriya Matematicheskaya 5 (1).

Kozintseva, A., 1999. Comparison of Three Methods of Spatial Prediction. thesis MSc. University of Maryland.

Krige, D.G., 1951. A statistical approach to some basic mine valuation problems on the Witwatersrand. Journal of the Chemical, Metallurgical and Mining Society of South Africa 119–139.

Krige, D.G., 1966. Two-dimensional weighted moving average trend surfaces for ore evaluation. Journal of the South African Institute of Mining and Metallurgy 66, 13–38.

Kumar, V., 2007. Optimal contour mapping of groundwater levels using universal kriging-a case study. Hydrological Sciences 52 (5), 1038–1050.

Lantuejoul, C., 2002. Geostatistical Simulation. Springer, New York.

Ling, M., Rifai, H.S., Newell, C.J., 2005. Optimizing groundwater long-term monitoring networks using Delaunay triangulation spatial analysis techniques. Environmetrics 16 (6), 635–657.

Ly, S., Charles, C., Degré, A., 2011. Geostatistical interpolation of daily rainfall at catchment scale: the use of several variogram models in the Ourthe and Ambleve catchments, Belgium. Hydrology and Earth System Sciences 15 (7), 2259–2274.

Matérn, B., 1986. In: Spatial Variation, second ed. Springer, Berlin.

Matheron, G., 1963. Principles of geostatistics. Economic Geology 1246–1266.

Matheron, G., 1969. Le Krigeage Universel (Universal Kriging), vol. 1. Ecole des Mines de Paris, Fontainebleau.

Matheron, G., 1971. The Theory of Regionalized Variables and its Applications. Ecole Nationale Superieure des Mines de Paris, Fontainebleau, Paris, p. 211.

Mendonca, C.A., Silva, J.B.C., 1995. Interpolation of potential-field data by equivalent layer and minimum curvature: a comparative analysis. Geophysics 60 (2), 399–407.

Moukana, J.A., Koike, K., 2008. Geostatistical model for correlating declining groundwater levels with changes in land cover detected from analyses of satellite images. Computers & Geosciences 34 (11), 1527–1540.

Mouser, P.J., Rizzo, D.M., Röling, W.F.M., Van Breukelen, B.M., 2005. A multivariate statistical approach to spatial representation of groundwater contamination using hydrochemistry and microbial community profiles. Environmental Science and Technology 39 (19), 7551–7559.

Mulchrone, K.F., 2003. Application of delaunay triangulation to the nearest neighbour method of strain analysis. Journal of Structural Geology 25 (5), 689–702.

Myers, D.E., 2005. Spatial interpolation comparison exercise 2004: a real problem or an academic exercise. In: Dubois, G. (Ed.), Automatic Mapping algorithms for Routine and Emergency Monitoring Data. EUR 21595 EN – Scientific and Technical Research Series. Office for Official Publications of the European Communities, Luxembourg, ISBN 92-894-9400-X, pp. *79–88*. EUR, 2005 ed.

Nelder, J.A., Mead, R., 1965. A simplex method for function minimization. Computer Journal 7 (4), 308–313.

Neuman, S., Jacobson, E., 1984. Analysis of nonintrinsic spatial variability by residual kriging with application to regional groundwater levels. Mathematical Geology 16 (5), 499–521.

Nikroo, L., Kompani-Zare, M., Sepaskhah, A., Fallah Shamsi, S., 2009. Groundwater depth and elevation interpolation by kriging methods in Mohr Basin of Fars province in Iran. Environmental Monitoring and Assessment 166 (1–4), 387–407.

Okabe, A., Boots, B., Sugihara, K., 1992. Spatial Tessellations: Concepts and Applications of Voronoi Diagrams. Wiley, New York.

Olea, R., 2006. A six-step practical approach to semivariogram modeling. Stochastic Environmental Research and Risk Assessment 20 (5), 307–318.

Olea, R.A., 1999. Geostatistics for Engineers and Earth Scientists. Kluwer Academic Publishers, New York.

Ouyang, Y., Zhang, J.E., Ou, L.T., 2006. Temporal and spatial distribution of sediment total organic carbon in an estuary river. Journal of Environmental Quality 35 (1), 93–100.

Özger, M., Şen, Z., 2007. Triple diagram method for the prediction of wave height and period. Ocean Engineering 34 (7), 1060–1068.

Pardo-Iguzquiza, E., Chica-Olmo, M., 2008. Geostatistics with the Matern semivariogram model: a library of computer programs for inference, kriging and simulation. Computers & Geosciences 34 (9), 1073–1079.

Pardo-Iguzquiza, E., Dowd, P., 2005. Empirical maximum likelihood Kriging: the general case. Mathematical Geology 37 (5), 477–492.

Philip, G.M., Watson, D.F., 1986. Automatic interpolation methods for mapping piezometric surfaces. Automatica 22 (6), 753–756.

Press, W.H., Teukolsky, S.A., Vettering, W.T., Flannery, B.P., 1992. Numerical Recipes in Fortran, second ed. Cambridge University Press, New York.

Pucci, A.A.J., Murashige, J.A.E., 1987. Applications of universal kriging to an aquifer study in New Jersey. Ground Water 25 (6), 672–678.

Rivest, M., Marcotte, D., Pasquier, P., 2008. Hydraulic head field estimation using kriging with an external drift: a way to consider conceptual model information. Journal of Hydrology 361 (3–4), 349–361.

Rivoirard, J., 2002. On the structural link between variables in Kriging with external drift. Mathematical Geology 34 (7), 797–808.

Rodriguez-Iturbe, I., Mejia, M.J., 1974. The design of rainfall networks in time and space. Water Resources Research 10 (4), 713–728.

Rouhani, S., 1986. Comparative study of ground-water mapping techniques. Ground Water 24 (2), 207–216.

Sahimi, M., 2011. Flow and Transport in Porous Media and Fractured Rock: From Classical Methods to Modern Approaches. Wiley-VCH.

Salas, J., 1993. Analysis and modeling of hydrologic time series. In: Maidment, D. (Ed.), Handbook of Hydrology, vol. 19. McGraw-Hill, New York, USA, pp. 11–19, 72.

Sandwell, D., 1987. Biharmonic spline interpolation of GEOS-3 and SEASAT altimeter Data. Geophysical Research Letters 14 (2), 139–142.

Schabenberger, O., Gotway, C.A., 2005. Statistical Methods for Spatial Data Analysis. CRC Press, Boca Raton, FL.

Sophocleous, M., Paschetto, J.E., Olea, R.A., 1982. Ground-water network design for northwest Kansas, using the theory of regionalized variables. Ground Water 20 (1), 48–58.

Stein, M.L., 1999. Interpolation of Spatial Data: Some Theory for Kriging. Springer, New York.

Subyani, A.M., Sen, Z., 1989. Geostatistical modelling of the Wasia aquifer in Central Saudi Arabia. Journal of Hydrology 110 (3–4), 295–314.

Sun, Y., Kang, S., Li, F., Zhang, L., 2009. Comparison of interpolation methods for depth to groundwater and its temporal and spatial variations in the Minqin oasis of northwest China. Environmental Modelling & Software 24 (10), 1163–1170.

Thyer, M., Kuczera, G., Wang, Q.J., 2002. Quantifying parameter uncertainty in stochastic models using the Box-Cox transformation. Journal of Hydrology 265 (1–4), 246–257.

Van den Boogaart, K.G., 2005. The comparison of one-click mapping procedures for emergency. In: Dubois, G. (Ed.), Automatic Mapping algorithms for Routine and Emergency Monitoring Data. EUR 21595 EN – Scientific and Technical Research Series. Office for official publications of the European Communities, Luxembourg, ISBN 92-894-9400-X, pp. *71–78*. EUR, 2005 ed.

Varouchakis, E.A., Hristopulos, D.T., 2013a. Comparison of stochastic and deterministic methods for mapping groundwater level spatial variability in sparsely monitored basins. Environmental Monitoring and Assessment 185 (1), 1–19.

Varouchakis, E.A., Hristopulos, D.T., 2013b. Improvement of groundwater level prediction in sparsely gauged basins using physical laws and local geographic features as auxiliary variables. Advances in Water Resources 52, 34–49.

Varouchakis, E.A., Hristopulos, D.T., Karatzas, G.P., 2012. Improving kriging of groundwater level data using nonlinear normalizing transformations-a field application. Hydrological Sciences Journal 57 (7), 1404–1419.

Wackernagel, H., 2003. Multivariate Geostatistics: An Introduction With Applications, third ed. Springer, Berlin.

Webster, R., Oliver, M., 2001. Geostatistics for Environmental Scientists: Statistics in Practice. Wiley, Chichester.

Wessel, P., 2009. A general-purpose Green's function-based interpolator. Computers & Geosciences 35 (6), 1247–1254.

Whittle, P., 1954. On stationary processes in the plane. Biometrika 41, 439–449.

Yilmaz, H.M., 2007. The effect of interpolation methods in surface definition: an experimental study. Earth Surface Processes and Landforms 32 (9), 1346–1361.

Zimmermann, B., Zehe, E., Hartmann, N.K., Elsenbeer, H., 2008. Analyzing spatial data: an assessment of assumptions, new methods, and uncertainty using soil hydraulic data. Water Resources Research 44 (10). https://doi.org/10.1029/2007WR006604.

Žukovič, M., Hristopulos, D.T., 2008. Environmental time series interpolation based on Spartan random processes. Atmospheric Environment 42 (33), 7669–7678.

Žukovič, M., Hristopulos, D.T., 2009. The method of normalized correlations: a fast parameter estimation method for random processes and isotropic random fields that focuses on short-range dependence. Technometrics 51 (2), 173–185.

Background of Spatiotemporal Geostatistical Analysis: Application to Aquifer Level Mapping

Emmanouil A. Varouchakis

SCHOOL OF ENVIRONMENTAL ENGINEERING, TECHNICAL UNIVERSITY OF CRETE, CHANIA, GREECE

1. Spatiotemporal Geostatistics

Space–time geostatistical approaches can be used to model the variability of environmental data. In areas with limited spatial and temporal data availability, application of space–time approaches can improve the reliability of predictions by incorporating space–time correlations instead of purely spatial ones; therefore the former approaches involve more parameters (Lee et al., 2010).

In Christakos (1991a,b) a theory of spatiotemporal random fields is developed and properties of the most important classes of spatiotemporal fields are examined. The theory is used to describe the correlation structure of space nonhomogeneous/time nonstationary processes and to derive optimal estimators for data dispersed simultaneously in space and time. Christakos and Hristopulos (1998) presented a completed review and new material on Bayesian maximum entropy estimation techniques and space–time random field estimation methods. Later, Kolovos et al. (2004) presented various methods for constructing space–time covariance models. These include nonseparable (in space and time) covariance models derived from physical laws (i.e., differential equations and dynamic rules), spectral functions, and generalized random fields. It is also shown that nonseparability is often a direct result of the physical laws that govern the process. The proposed methods can generate covariance models for homogeneous/stationary as well as for nonhomogeneous/nonstationary environmental processes across space and time.

Kyriakidis and Journel (1999) presented an extensive review for space–time geostatistical techniques. The initial approach for space–time geostatistical analysis was to add time as an additional dimension of space (Kyriakidis and Journel, 1999; Rouhani and Myers, 1990). This approach was implemented using the kriging technique developing space–time kriging. Advanced space–time geostatistical approaches were also

Spatiotemporal Analysis of Extreme Hydrological Events. https://doi.org/10.1016/B978-0-12-811689-0.00002-1

developed by Christakos (1991b, 2000) and Kyriakidis and Journel (1999) to account for fundamental dependencies in the combined space–time metric (Lee et al., 2010).

Bayesian approaches as an alternative to non-Bayesian, i.e., kriging, were introduced by Christakos (1990, 2000). Bayesian Maximum Entropy (BME) is a nonlinear method that relies on a two-step procedure that first involves a Maximum Entropy step (the ME part of BME) to obtain a prior distribution and on a Bayesian conditioning rule for the assimilation of secondary information (possibly soft data). BME provides a flexible framework that accounts for the wide variety of available knowledge bases and leads, in general, to optimal nonlinear space–time estimators. Applications include soil science, e.g., Brus et al. (2008), water consumption (Lee and Wentz, 2008), environmental health studies, e.g., Christakos and Hristopulos (1998), Kolovos et al. (2012), and Yu et al. (2009), and atmospheric environment, e.g., Christakos and Serre (2000), Vyas and Christakos (1997), and Yu et al. (2011).

A framework for stochastic spatiotemporal modeling has also been presented by Kyriakidis (2001a) and Kyriakidis and Journel (2001b). A dataset that is more densely sampled in time than in space can be modeled via a set of spatially correlated time series (Rouhani and Hall, 1989). The time series at each sampled location can be decomposed into a nonstationary deterministic or stochastic trend component and a stationary residual component. The residual time series is then fitted with a covariance model. It is possible to apply this approach to perform spatial interpolation or extrapolation; extending it to a continuous spatial domain by determining temporal covariance models or time series independently at each fixed location and then regionalizing them in space. Time series regionalization involves simulation of the spatiotemporal residual field by generating simulated realizations at any unmonitored location: sequential Gaussian simulation, i.e., autoregression (Kyriakidis and Journel, 2001a). Simple kriging (SK) is used for covariance parameter regionalization. This allows temporal covariance models or time series to be determined at unsampled locations and reduces the computational effort associated with the number of (temporal) covariances. A simulation procedure is also used for trend regionalization, which is typically approximated by periodic and sine and cosine functions in conjunction with multiple regression. The independently simulated trend and residual realizations are then added to produce realizations for the spatiotemporal field. An estimate of the standard deviation of the unknown residual profile at any unmonitored location is also derived via SK. Although this framework has been characterized as powerful, it requires multiple regionalizations, thus time and computational load (Kyriakidis and Journel, 2001b).

Space–time kriging has been applied in geohydrology by Rouhani and Hall (1989) where intrinsic random functions (polynomial spatiotemporal covariance) for space–time kriging of piezometric data were used. In Rouhani and Myers (1990) potential drawbacks of space–time geostatistical analysis on geohydrological data (piezometric data) are discussed. More recently, space–time kriging was applied successfully in groundwater level spatiotemporal modeling using monthly and biannual datasets (Varouchakis, 2017; Varouchakis and Hristopulos, 2017). In addition, Mendoza-Cazares and Herrera-Zamarron (2010) applied space–time kriging for the estimation of the water level of the Queretaro-Obrajuelo aquifer, while Hoogland et al. (2010) mapped the seasonal fluctuation of water-table depths in Dutch nature conservation areas. Furthermore, space–time kriging was used for the design of rainfall networks in time and space (Rodriguez-Iturbe and Mejia, 1974) and in a comparison study for estimating runoff time series in ungauged locations (Skøien and Blöschl, 2007).

Space–time kriging has also been used in a wide range of scientific fields and topics such as agriculture (Heuvelink and Egmond, 2010; Stein, 1998), atmospheric data (De Iaco et al., 2002b; Myers, 2002; Nunes and Soares, 2005), soil science water content (Jost et al., 2005; Snepvangers et al., 2003), surface temperature data (Hengl et al., 2011), wind data (Gneiting, 2002), gamma radiation data (Heuvelink and Griffith, 2010), epidemiology (Gething et al., 2007), and forecasting municipal water demand (Lee et al., 2010).

Space–time geostatistical analysis is based on the joint spatial and temporal dependence between observations. There are two ways to represent space–time random variables (Christakos, 1991b): (1) full space–time models using separable or non-separable space–time covariance functions $Z(\mathbf{s}, t)$, $(\mathbf{s}, t) \in D \times T$, where $D \subseteq \mathrm{R}^d$ is the spatial domain (d is the dimensions) and $T \subseteq \mathrm{R}$ is the temporal domain, and (2) vectors of temporally correlated spatial random fields $Z(\mathbf{s}, t) = Z_t(\mathbf{s})$, $t = 1 \ldots T$, where T is the number of temporally correlated spatial random fields or vectors of spatially correlated time series $Z(\mathbf{s}, t) = Z_{\mathbf{s}}(t)$, $\mathbf{s} = 1 \ldots n$, where n is the number of locations. The representation depends on the domain density (space or time).

The space–time kriging method employs the first model. The two main tasks of space–time analysis are interpolation and extrapolation. The first refers to estimation of variable values at unmeasured locations inside the spatial extent of the study area, while the latter extends the estimations ahead of the boundaries of the observations in space or time. The main assumption used in interpolation and extrapolation is that the specific patterns extracted from the available data analysis delivers sufficient information to capture the spatiotemporal dynamics of the observed data (Lee et al., 2010).

The application of space–time kriging to space–time field data entails practical difficulties, especially for geohydrological data. The most important problem is the construction of valid covariance or variogram models in space–time; valid covariance or variogram models constructed in the spatial or temporal context are not, in general, valid when a valid temporal model with a valid spatial model are combined to produce a spatiotemporal model. Geohydrological data are usually dense in time and sparse in space. This feature is significant since covariances or variograms can lead to significantly different levels of reliability in space and time. The kriging estimator is inferior if the data are collected during the wet season and the estimates refer to the dry season. Finally, in space–time kriging applications, computational problems may arise for specific sampling patterns with the coefficient matrix in the kriging system. These problems are due, e.g., to insufficient numbers of sample locations compared to the order of a drift function applied to the data, or scarcity and clustering of sampling locations (Rouhani and Myers, 1990).

2. Spatiotemporal Geostatistical Modeling

Spatiotemporal geostatistical models provide a probabilistic framework for data analysis and predictions, which is based on the joint spatial and temporal dependence between observations (Fischer and Getis, 2010; Kyriakidis and Journel, 1999). Initial approaches to spatiotemporal data modeling were based on separable covariance functions, obtained by combining separate spatial and temporal covariance models (Cressie, 1993; Dimitrakopoulos and Luo, 1994; Rodriguez-Iturbe and Mejia, 1974; Rouhani and Myers, 1990). The last two decades have seen significant development of nonseparable covariance functions. These models aim to improve spatiotemporal data modeling and prediction (Cressie and Huang, 1999; De Iaco et al., 2001; Gneiting, 2002; Kolovos et al., 2004) by extracting in some cases the covariance functions from physical laws such as differential equations and dynamic rules (Christakos, 2000; Christakos and Hristopulos, 1998; Gneiting, 2002; Kolovos et al., 2004).

The main goal of space–time analysis is to model multiple time series of data at spatial locations where a distinct time series is allocated. The time variable is considered as an additional dimension in geostatistical prediction. A spatiotemporal stochastic process can be represented by $Z(\mathbf{s}, t)$ where the variable of interest of random field Z is observed at N space–time coordinates $(\mathbf{s}_i, t_i), \ldots (\mathbf{s}_N, t_N)$, while the optimal prediction of the variable in space and time is based on $Z(\mathbf{s}_i, t_i), \ldots Z(\mathbf{s}_N, t_N)$ (Cressie and Huang, 1999; Giraldo Henao, 2009). Space–time random fields (S/TRF) $Z(\mathbf{s}, t)$ can be decomposed into a mean component $m_Z(\mathbf{s}, t)$ modeling the trend and a residual S/TRF component $Z'(\mathbf{s}, t)$ modeling fluctuations around that trend in both space and time according to the following equation:

$$Z(\mathbf{s}, t) = m_Z(\mathbf{s}, t) + Z'(\mathbf{s}, t). \tag{2.1}$$

The trend term can be calculated using a deterministic method while, the fluctuations using a stochastic framework such as space–time kriging (Christakos, 1991b; Kyriakidis and Journel, 1999).

2.1 Spatiotemporal Two-Point Function

Set $Z(\mathbf{s}, t)$, $(\mathbf{s}, t) \in D \times T$, a second-order stationary space–time random field. $D \subseteq \mathrm{R}^d$ is the spatial domain (d is the space dimensions) and $T \subseteq \mathrm{R}$ is the temporal domain, with expected value (Myers et al., 2002): $E[Z(\mathbf{s}, t)] = 0$, $\forall (\mathbf{s}, t) \in D \times T$, and covariance function:

$$C_{ST}(\mathbf{r}_s, r_t) = E[Z(\mathbf{s}_j + \mathbf{r}_s,\ t_j + r_t)\, Z(\mathbf{s}_i, t_i)] - E[Z(\mathbf{s}_j + \mathbf{r}_s,\ t_j + r_t)]E[Z(\mathbf{s}_i, t_i)], \tag{2.2}$$

where $\mathbf{r}_s = (\mathbf{s}_i - \mathbf{s}_j)$, $r_t = (t_i - t_j)$, $i, j = 1, \ldots, N$. The covariance function depends only on the lag vector $\mathbf{r} = (\mathbf{r}_s, r_t)$ and not on location or time, while it must satisfy the positive-definiteness condition to be a valid covariance function. Hence for any $(\mathbf{s}_i, t_i) \in D \times T$, any real a_i, $i = 1, \ldots, N$, and any positive integer N, C_{ST} must satisfy the following inequality[1]:

$$\sum_{i=1}^{N}\sum_{j=1}^{N} a_i a_j C_{ST}(\mathbf{s}_i - \mathbf{s}_j, t_i - t_j) > 0.$$

If $E[Z(\mathbf{s}, t)]$ is constant and $C_{ST}(\mathbf{r}_s, r_t)$ depends only on the lag vector $\mathbf{r} = (\mathbf{r}_s, r_t)$:

$$\mathrm{Cov}(Z(\mathbf{s}_i, \mathbf{s}_j; t_i, t_j)) = C_{ST}(\mathbf{s}_i - \mathbf{s}_j, t_i - t_j) = C_{ST}(\mathbf{r}_s, r_t). \tag{2.3}$$

the S/TRF $Z(\mathbf{s}, t)$ is characterized as *second-order stationary*. Spatial and spatiotemporal geostatistical prediction methodologies generally rely on stationarity (stationary mean and covariance or variogram).

In addition the field is isotropic if:

$$C_{ST}(\mathbf{r}_s, r_t) = C_{ST}(\|\mathbf{r}_s\|, |r_t|), \tag{2.4}$$

meaning that the covariance function depends only on the length of the lag.

Under the weaker intrinsic stationarity assumption the increment $(Z(\mathbf{s}_j + \mathbf{r}_s,\ t_j + r_t) - Z(\mathbf{s}_i, t_i))$ is second-order stationary for every lag vector $\mathbf{r}_s, r_t$ instead of the random field. Then $Z(\mathbf{s}, t)$ is called an intrinsic random function and is characterized by:

$$E(Z(\mathbf{s}_j + \mathbf{r}_s,\ t_j + r_t) - Z(\mathbf{s}_i, t_i)) = 0, \tag{2.5}$$

and:

$$\gamma_{ST}(\mathbf{r}_s, r_t) = \frac{1}{2}\mathrm{var}(Z(\mathbf{s}_j + \mathbf{r}_s,\ t_j + r_t) - Z(\mathbf{s}_i, t_i)) \tag{2.6}$$

where the term var denotes the variance. The function $\gamma_{ST}(\mathbf{r}_s, r_t)$ only depends on the lag vector $\mathbf{r} = (\mathbf{r}_s, r_t)$. The quantity $\frac{1}{2}\mathrm{var}(Z(\mathbf{s}_j + \mathbf{r}_s,\ t_j + r_t) - Z(\mathbf{s}_i, t_i))$ is called the semi-variance at lag $\mathbf{r} = (\mathbf{r}_s, r_t)$.

The random field $Z(\mathbf{s}, t)$ has an intrinsically stationary variogram if it is intrinsically stationary with respect to both space and time dimensions. The $Z(\mathbf{s}, t)$ has a spatially intrinsically stationary variogram if the variogram depends only on the spatial separation vector $\mathbf{r}_s$ for every pair of time instants t_i, t_j and it has a temporally intrinsically stationary variogram if it depends only on the temporal lag r_t. Eq. (2.6) provides the space–time stationary variogram function (Gneiting et al., 2007). Under the stronger assumption of second-order stationarity, the semivariance is defined as:

$$\gamma_{ST}(\mathbf{r}_s, r_t) = C_{ST}(0,0) - C_{ST}(\mathbf{r}_s, r_t). \tag{2.7}$$

The primary concerns when modeling space–time structures are to ensure that the chosen model is valid and that the model is suitable for the data. The space–time kriging estimator can be applied if the space–time covariance function satisfies the positive-definiteness condition, $C_{ST} > 0$, explained earlier (Cressie and Huang, 1999). The model's suitability is ensured by testing a series of available structures on the data. The variogram function must be conditionally negative definite to ensure that the space–time kriging equations have a valid unique solution (De Iaco, 2010; Myers et al., 2002).

[1]Positive-definiteness condition is often also presented as nonnegative-definiteness condition, i.e., ≥ 0.

Space–time kriging is a well-established method for space–time interpolation (Christakos et al., 2001; De Cesare et al., 2001). It is, however, complicated, because the kriging system of equations needs to be solved at the same time for spatial and temporal weights (Skøien and Blöschl, 2007). In addition, space–time kriging is data demanding, while often the number of locations where time series of groundwater level data are available is very limited. Also according to Bierkens (2001), space–time kriging may not be appropriate to analyze the change in groundwater level if climate change effects (rainfall shortage, intense rainfall periods, droughts) affect the area of study or changes of land use and surface water management occur. The kriging estimator and the kriging equations have the same form for spatiotemporal problems as for spatial problems. The difference from spatial-only kriging is the covariance modeling. This is because the time component is not an extra dimension that can be used to form a single Euclidean space–time metric. The time axis is by nature different and not necessarily orthogonal to the three spatial axes. The time component has been proved to cause both theoretical and practical problems if it is treated as an additional space dimension (Rouhani and Myers, 1990). Hence in space–time variography the spatial lag, $\mathbf{r}_s$, and the temporal lag, r_t, are treated as independent arguments $\gamma(\mathbf{r}_s;r_t)$. Space–time and purely spatial kriging methods were analytically presented and compared on simulated data by Bogaert (1996). He concludes that in the space–time context, ordinary space–time kriging is preferable; nevertheless, it requires the hypothesis of mean and variance homogeneity and is limited to second-order stationary random fields.

Two categories of models are used for variogram or covariance modeling. The first includes separable models whose covariance function is a combination of a spatial and a temporal covariance function; the second includes nonseparable models in which a single function models the spatiotemporal data dependence. Separable models, however, suffer from unrealistic assumptions and properties (Hengl et al., 2011; Snepvangers et al., 2003). Both space–time covariance models are valid according to De Iaco et al. (2001, 2002b) and Cressie and Huang (1999).

Separable and nonseparable covariance functions can describe the random field's spatiotemporal continuity. Separable covariance functions are used if separate spatial and separate temporal covariance functions exist for the data (Gneiting et al., 2007):

$$\mathrm{Cov}(Z(\mathbf{s}_i,\mathbf{s}_j;t_i,t_j)) = C_{ST}(\mathbf{r}_s,r_t) = C_{\mathbf{r}_s}(\mathbf{s}_i,\mathbf{s}_j)\ C_{r_t}(t_i,t_j). \tag{2.8}$$

Separability provides many advantages, such as the simplified representation of the covariance matrix and consequently important computational benefits (Park and Fuentes, 2008). The separable covariance models, however, in spite of their simplicity are not usually physically motivated. Correlations that have separable spatial and temporal components are particularly useful when the correlations are inferred on the basis of existing data and not when they follow from a physical model (Christakos and Hristopulos, 1998). When data (e.g., hydrologic, atmospheric, oceanographic) are influenced by dynamic processes, spatiotemporal dependency structures are difficult to model by a separable covariance function. Physically meaningful covariance models can be derived instead, based on environmental data dynamic processes (Christakos, 1991b; Christakos and Hristopulos, 1998; Gneiting, 2002; Kolovos et al., 2004). Covariance structures such these are nonseparable. Modeling nonseparable covariance functions is one of the keys for more reliable prediction in the environmental research fields (Gneiting et al., 2007).

The random field $Z(\mathbf{s}, t)$ has fully symmetric separable covariance if (De Iaco, 2010; Gneiting et al., 2007):

$$\mathrm{Cov}(Z(\mathbf{s}_i, t_i), Z(\mathbf{s}_j, t_j)) = \mathrm{Cov}(Z(\mathbf{s}_i, t_j), Z(\mathbf{s}_j, t_i)), \tag{2.9}$$

for all $(\mathbf{s}_i, t_i)$, $(\mathbf{s}_j, t_j)$. Nonseparable covariance structures that are not fully symmetric have been proposed by Gneiting et al. (2007), e.g.:

$$C_{ST}(\mathbf{r}_s, r_t) = (1 + r_t)^{-1}\exp\left[\mathbf{r}_s \big/ (1 + r_t)^{\beta/2}\right] \quad 0 \le \beta \le 1. \tag{2.10}$$

A significant part in the space–time process is the choice of the variogram or covariance model and the estimation of its parameters. Contrary to purely spatial prediction, where a well-established set of variogram models exists, several spatiotemporal models have been developed for modeling space–time structures (Christakos and Hristopulos, 1998; De Cesare et al., 2001; Gething et al., 2007; Kyriakidis and Journel, 1999). These models involve the product model (Rodriguez-Iturbe and Mejia, 1974), the sum model (Rouhani and Hall, 1989), the metric model (Dimitrakopoulos and Luo, 1994), the integrated product model (Cressie and Huang, 1999; De Iaco et al., 2002a), the product–sum model (De Cesare et al. 2001, 2002), the integrated product–sum model (De Iaco et al. 2002a, 2002b), Gneiting's nonseparable models (Gneiting, 2002; Gneiting et al., 2007), a series of nonseparable models reviewed in Kolovos et al. (2004), and nonseparable models expressed through the spectrum density function instead of the direct covariance function (Porcu et al., 2008).

2.2 Spatiotemporal Covariance or Variogram Models

A comprehensive description of some widely used spatiotemporal covariance or variogram models follows.

The metric model is given by the following equation (Dimitrakopoulos and Luo, 1994):

$$C_{ST}(\mathbf{r}_s, r_t) = C\left(\alpha_1 \|\mathbf{r}_s\|^2 + \alpha_2 |r_t|^2\right), \tag{2.11}$$

or:

$$\gamma_{ST}(\mathbf{r}_s, r_t) = \gamma\left(\alpha_1 \|\mathbf{r}_s\|^2 + \alpha_2 |r_t|^2\right), \tag{2.12}$$

where $\|\mathbf{r}_s\|$ is the Euclidean norm of the spatial lag vector and α_1, α_2 are coefficients that weigh relatively the space and time contributions. For this model the spatial and temporal covariances used are of the same type.

Another separable space–time covariance model is the sum model, in which spatial $C_S(\mathbf{r}_s)$ and temporal $C_T(r_t)$ covariance functions are added (Rouhani and Hall, 1989):

$$C_{ST}(\mathbf{r}_s, r_t) = C_S(\mathbf{r}_s) + C_T(r_t), \tag{2.13}$$

or:

$$\gamma_{ST}(\mathbf{r}_s, r_t) = \gamma_S(\mathbf{r}_s) + \gamma_T(r_t). \tag{2.14}$$

where C_{ST} is the spatiotemporal covariance and γ_{ST} is the spatiotemporal variogram, respectively. According to Rouhani and Myers (1990), covariance matrices $C_{ST}(\mathbf{r}_s, r_t)$ of certain configurations of space–time data can be singular. In this case the covariance function is only positive semidefinite $C_{ST} \geq 0$ (De Iaco, 2010). The sum expression therefore is almost an acceptable model as it only fails the strict-definiteness condition. The resulting spatial–temporal form of covariance or variogram does not satisfy the strict-definiteness conditions for the separate spatial and temporal covariances and the strict conditional negative-definiteness condition for the separate spatial and temporal variograms (Dimitrakopoulos and Luo, 1994; Myers and Journel, 1990; Rouhani and Myers, 1990). Thus this model is unsatisfactory for optimal prediction (De Iaco, 2010).

The product model (Rodriguez-Iturbe and Mejia, 1974) belongs to the separate space–time model category and is one of the simplest ways to model a covariance or variogram in space–time. The product of a space variogram and a time variogram is generally not a valid variogram; on the other hand, the product of a space covariance and a time covariance leads to a valid model. A variogram structure can then be determined by the product covariance model. Valid spatial and temporal covariance models can be used in the following product form to create spatiotemporal models:

$$C_{ST}(\mathbf{r}_s, r_t) = C_S(\mathbf{r}_s)C_T(r_t). \tag{2.15}$$

If both components $C_S(\mathbf{r}_s), C_T(r_t)$ are strictly positive definite, then $C_{ST}(\mathbf{r}_s, r_t)$ is strictly positive definite on $\mathrm{R}^d \times T$. The covariance equation can be expressed in terms of the variogram as:

$$\gamma_{ST}(\mathbf{r}_s, r_t) = C_T(0)\gamma_S(\mathbf{r}_s) + C_S(0)\gamma_T(r_t) - \gamma_S(\mathbf{r}_s)\gamma_T(r_t). \tag{2.16}$$

The product–sum space–time model (De Cesare et al. 2001, 2002) is a generalization of the product and the sum model, while it constitutes the starting point for its integrated product sum versions. It is defined as:

$$C_{ST}(\mathbf{r}_s, r_t) = k_1 C_S(\mathbf{r}_s)C_T(r_t) + k_2 C_S(\mathbf{r}_s) + k_3 C_T(r_t). \tag{2.17}$$

where C_S, C_T are purely spatial and temporal covariance models with $k_1 > 0, k_2 \geq 0, k_3 \geq 0$. If $C_S(\mathbf{r}_s)$ and $C_T(r_t)$ are strictly positive definite, then $C_{ST}(\mathbf{r}_s, r_t)$ is strictly positive definite on $\mathrm{R}^d \times T$. In terms of the variogram, the foregoing equation is expressed as:

$$\gamma_{ST}(\mathbf{r}_s, r_t) = (k_1 C_S(0) + k_3)\gamma_T(r_t) + (k_1 C_T(0) + k_2)\gamma_S(\mathbf{r}_s) - k_1\gamma_S(\mathbf{r}_s)\gamma_T(r_t), \tag{2.18}$$

where γ_S, γ_T are purely spatial and temporal variogram models. $C_S(0)$ and $C_T(0)$ are the sills of the spatial and temporal variograms, respectively. Each space–time model (sum, product) have separate limitations, which their combination does not have. The variogram structure can be expressed alternatively as follows (De Iaco et al., 2001):

$$\gamma_{ST}(\mathbf{r}_s, r_t) = \gamma_{ST}(\mathbf{r}_s, 0) + \gamma_{ST}(0, r_t) - K\gamma_{ST}(\mathbf{r}_s, 0)\gamma_{ST}(0, r_t), \tag{2.19}$$

where $0 < \mathrm{K} \leq 1/\max(\text{sill } \gamma_{ST}(\mathbf{r}_s, 0)\ , \text{ sill } \gamma_{ST}(0, r_t))$.

The Cressie–Huang models (Cressie and Huang, 1999) are nonseparable spatiotemporal stationary covariance functions defined by:

$$C_{ST}(\mathbf{r}_s, r_t) = \int e^{i\mathbf{r}_s^T\omega}\rho(\omega, r_t)k(\omega)d\omega, \tag{2.20}$$

where $\rho(\omega, \cdot)$ is a continuous autocorrelation function $\forall\, \omega \in \mathrm{R}^d$ and $k(\cdot)$ is a positive function with $k(\omega) > 0$ and $\int k(\omega)d\omega < \infty$. Bochner's theorem is used to derive nonseparable space–time covariance functions of this type.

Gneiting (2002) proposed a wide class of nonseparable covariances derived from the following equation:

$$C_{ST}(r_s, r_t) = \frac{\sigma^2}{\left[\psi(r_t^2)\right]^{d/2}}\varphi\left(\frac{\|\mathbf{r}_s\|^2}{\psi(r_t^2)}\right) \quad (\mathbf{r}_s, r_t) \in \mathrm{R}^d \times T, \tag{2.21}$$

where d is the number of spatial dimensions, $\varphi(\tau)$, $\tau \geq 0$, is a completely monotone function, and $\psi(\tau)$, $\tau \geq 0$, is a positive function (i.e., Bernstein function or equivalently a variogram) with a completely monotone derivative. A real and positive function $f : [0, \infty] \rightarrow [0, \infty]$ is called completely monotone if and only if $(-1)^N f^{(N)}(\tau) \geq 0$, for any positive integer N (Porcu et al., 2006). Examples of such functions are given in Gneiting (2002). The spatial and temporal structures are determined by φ and ψ, respectively. However, contrary to the Cressie–Huang models the Gneiting models do not recall Bochner's theorem.

A similar approach to the Cressie–Huang models can be formulated for the product (Eq. 2.15) and product-sum (Eq. 2.17) constructions. Their integration also gives valid spatiotemporal models (De Iaco et al., 2002a; Myers et al., 2002) as follows:

$$C_{ST}(\mathbf{r}_s, r_t) = \int_V kC_S(\mathbf{r}_s; a)C_T(r_t; a)\, \mathrm{d}\mu(a), \tag{2.22}$$

where $k > 0$, $\mu(a)$ is a positive measure on $U \subseteq \mathrm{R}$, $C_S(\mathbf{r}_s; a)$, $C_T(r_t; a)$ are valid covariance functions in $D \subseteq \mathrm{R}^d$ and $T \subseteq \mathrm{R}$, respectively, for each $a \in V \subseteq U$ and $C_S(\mathbf{r}_s; a)C_T(r_t; a)$ is integrable with respect to the measure μ on V for all $\mathbf{r}_s$, r_t. The integrated product model generates nonseparable and nonintegrable models.

In terms of the variogram structure the foregoing equation is rewritten as:

$$\gamma_{ST}(\mathbf{r}_s, r_t) = \int_V k[C_T(0; a)\gamma_S(\mathbf{r}_s; a) + C_S(0; a)\gamma_T(r_t; a). \\ -\gamma_S(\mathbf{r}_s; a)\gamma_T(r_t; a)]\mathrm{d}\mu(a) \tag{2.23}$$

Similarly for the product–sum model one obtains:

$$C_{ST}(\mathbf{r}_s, r_t) = \int_V [k_1 C_S(\mathbf{r}_s; a) C_T(r_t; a) + k_2 C_S(\mathbf{r}_s; a) + k_3 C_T(r_t; a)] \mathrm{d}\mu(a), \tag{2.24}$$

where $k_1 C_S(\mathbf{r}_s; a) C_T(r_t; a) + k_2 C_S(\mathbf{r}_s; a) + k_3 C_T(r_t; a)$ is integrable with respect to the measure μ on V for all $\mathbf{r}_s, r_t$ given $k_1 > 0$, $k_2 \geq 0$, $k_3 \geq 0$. Separate structures in space and time are used to generate the product–sum model and integrated product–sum model. In addition, the integrated product–sum model (Eq. 2.24) and the product–sum model (Eq. 2.17) are nonintegrable with respect to $\mathbf{r}_s$ and r_t and nonseparable. Eq. (2.24) can be written in terms of a variogram structure as:

$$\begin{aligned} \gamma_{ST}(\mathbf{r}_s, r_t) = \int_V & [(k_1 C_S(0; a) + k_3) \gamma_T(r_t; a) + (k_1 C_T(0; a) + k_2) \gamma_s(\mathbf{r}_s; a), \\ & - k_1 \gamma_S(\mathbf{r}_s; a) \gamma_T(r_t; a)] d\mu(a) \end{aligned} \tag{2.25}$$

where $C_S(0; a)$ and $C_T(0; a)$ are the corresponding sill values of the spatial and temporal variograms. Both space–time variogram structures are valid if $\gamma_S(\mathbf{r}_s; a)$ and $\gamma_T(r_t; a)$ are valid spatial and temporal variogram models.

2.3 Spatiotemporal Models' Summary of Characteristics

The metric model in spite of its nice asymptotic features has restrictive assumptions. As previously mentioned the same type of covariances describe the spatial and temporal correlation, they have the same sill (if the model is bounded) and it can be used only for processes whose space–time correlation is described by a model with geometric anisotropy. Finally, it is the only model that requires a space–time metric (De Iaco, 2010).

The product model, the product–sum model, their integrated versions, and the sum model are produced by separate space and time functions. The main advantage of such models is their ease of use in modeling and estimation. Because the sum model is separable, anisotropy can be incorporated in the spatial component. The product model is separable and integrable; its integrated version can generate nonseparable and nonintegrable models. In contrast the product–sum model is nonintegrable with respect to $\mathbf{r}_s$ and r_t, and it is nonseparable as the integrated version of the product–sum model. On the other hand, the Cressie–Huang and Gneiting models are alternative choices to the separable models. However, the φ and ψ functions in the Gneiting model can be chosen so that separable models are obtained. Finally, anisotropic covariance or variogram versions of the functions described can be constructed by inserting anisotropy in the spatial component of the variogram function (De Iaco, 2010; Myers et al., 2002).

The space–time variogram models described previously, except for the metric function, are typically used to model the space–time experimental variogram because an arbitrary space–time metric is not required and the fitting process is similar to that for spatial variograms (De Iaco, 2010; Gething et al., 2007).

2.4 A Spatiotemporal Covariance Function Derived From a Physical Law

An alternative covariance function suitable for space–time variogram modeling is proposed by Christakos and Hristopulos (1998) and Kolovos et al. (2004). It is an extension of a nonseparable spatiotemporal covariance inspired from the diffusion equation. A covariance model derived from a physical differential equation, such as the diffusion equation, is of the form (Christakos, 2000):

$$C_{ST}(\mathbf{r}_s, r_t) = (4\alpha\pi|r_t|)^{-n/2}\exp(-\mathbf{r}_s^2/4\alpha|r_t|) \quad \alpha > 0. \tag{2.26}$$

An extension of this equation can be obtained following Gneiting's proposition to add constants after the time lag in space–time formulations. Hristopulos (2002) proposed a similar approach for the spatial-only case. Therefore the equation can be modified to:

$$C_{ST}(\mathbf{r}_s, r_t) \cong (\beta r_t^{2\gamma} + 1)^{-n/2}\exp(-\mathbf{r}_s^2/(\beta r_t^{2\gamma} + 1)) \quad 0 \le \beta \le 1,\ 0 < \gamma \le 1, \tag{2.27}$$

where β, γ are the function's parameters and n are the dimensions. This covariance class has been used in the area of fluid mechanics (Monin and Yaglom, 1975).

2.5 Spatiotemporal Geostatistical Analysis and Prediction

Under the second-order stationarity hypothesis, the variogram and the covariance function are equivalent. For reasons of convenience the variogram structure is preferred. The appropriate variogram structure (separable or nonseparable) is fitted to the experimental spatiotemporal model given by:

$$\widehat{\gamma}_Z(\mathbf{r}_s, r_t) = \frac{1}{2N(\mathbf{r}_s, r_t)}\sum_{N(\mathbf{r}_s, r_t)}[Z(\mathbf{s}_i, t_i) - Z(\mathbf{s}_j + \mathbf{r}_s, t_j + r_t)]^2, \tag{2.28}$$

where $\mathbf{r}_s = ||\mathbf{s}_i - \mathbf{s}_j||$, $r_t = |t_i - t_j|$, and $N(\mathbf{r}_s, r_t)$ is the number of pairs in $N(\mathbf{r}_s, r_t)$. The space–time experimental variogram is estimated as half the mean squared difference between data separated by a given spatial and temporal lag $(\mathbf{r}_s, r_t)$.

Geostatistical prediction is then achieved using space–time ordinary kriging (STOK) (Christakos, 1991b; Goovaerts, 1997). The STOK estimator with respect to residual data notation is:

$$\widehat{Z}'(\mathbf{s}_0, t_0) = \sum_{\{i:\mathbf{s}_i, t_i \in \mathbb{S}_0\}} \lambda_i Z'(\mathbf{s}_i, t_i), \tag{2.29}$$

where $\widehat{Z}'(\mathbf{s}_0, t_0)$ is the unsampled location–time, $Z'(\mathbf{s}_i, t_i)$ are the sampled location–time neighbors, and λ_i are the corresponding space–time kriging weights:

$$\sum_{\{i:\mathbf{s}_i, t_i \in \mathbb{S}_0\}} \lambda_i \gamma_{Z'}(\mathbf{s}_i, \mathbf{s}_j; t_i, t_j) + \mu = \gamma_{Z'}(\mathbf{s}_j, \mathbf{s}_0; t_j, t_0), \quad j = 1, \ldots, N_0, \tag{2.30}$$

$$\sum_{\{i:\mathbf{s}_i, t_i \in \mathbb{S}_0\}} \lambda_i = 1, \tag{2.31}$$

where N_0 is the number of points within the search neighborhood of $\mathbf{s}_0$, $\gamma_{Z'}(\mathbf{s}_i, \mathbf{s}_j; t_i, t_j)$ is the variogram between two sampled points $\mathbf{s}_i$ and $\mathbf{s}_j$ at times t_i, and t_j, $\gamma_{Z'}(\mathbf{s}_j, \mathbf{s}_0; t_j, t_0)$ is the variogram between $\mathbf{s}_j$, t_j and the estimation point $\mathbf{s}_0$, t_0, and μ is the Lagrange multiplier enforcing the zero bias constraint.

The STOK estimation variance is given by the following equation, with the Lagrange coefficient μ compensating for the uncertainty of the mean value:

$$\sigma_E^2(\mathbf{s}_0, t_0) = \sum_{\{i:\mathbf{s}_i, t_i \in \mathbb{S}_0\}} \lambda_i \gamma_{Z'}(\mathbf{s}_j, \mathbf{s}_0; t_j, t_0) + \mu. \tag{2.32}$$

The prediction is also described in matrix notation where the system $\mathbf{\Gamma}\ \boldsymbol{\lambda} = \mathbf{c}$ is solved to estimate the spatiotemporal weights $\boldsymbol{\lambda}$:

$$\mathbf{\Gamma} = \begin{pmatrix} \widehat{\gamma}_{Z'}(\mathbf{s}_1 - \mathbf{s}_1, t_1 - t_1) & \cdots & \widehat{\gamma}_{Z'}(\mathbf{s}_1 - \mathbf{s}_{N_0}, t_1 - t_{N_0}) & 1 \\ \vdots & \ddots & \vdots & \\ \widehat{\gamma}_{Z'}(\mathbf{s}_{N_0} - \mathbf{s}_1, t_{N_0} - t_1) & \cdots & \widehat{\gamma}_{Z'}(\mathbf{s}_{N_0} - \mathbf{s}_{N_0}, t_{N_0} - t_{N_0}) & 1 \\ 1 & \cdots & 1 & 0 \end{pmatrix}$$

$$\boldsymbol{\lambda} = \begin{pmatrix} \lambda_1 \\ \vdots \\ \lambda_{N_0} \\ \mu \end{pmatrix}, \tag{2.33}$$

$$\mathbf{c} = \begin{pmatrix} \widehat{\gamma}_{Z'}(\mathbf{s}_0 - \mathbf{s}_1, t_0 - t_1) \\ \vdots \\ \widehat{\gamma}_{Z'}(\mathbf{s}_0 - \mathbf{s}_{N_0}, t_0 - t_{N_0}) \\ 1 \end{pmatrix}$$

where $\mathbf{\Gamma}$ is the matrix of the spatiotemporal variogram between the observed space–time data locations, $\boldsymbol{\lambda}$ are the spatiotemporal weights, and $\mathbf{c}$ is the matrix of the spatiotemporal variogram between the observed space–time data locations and the space–time estimation location.

Space–time predictions are usually based on a space–time neighborhood that encloses observations inside a search radius in space and time; the search radii depend on the space and time correlation lengths ξ_s, ξ_t estimated from the variogram fitting process. For small datasets the entire dataset is used for predictions. Fig. 2.1 presents a schematic representation of the space–time domain and the space–time search neighborhood.

In spatiotemporal regression kriging the estimate of the head, groundwater level, is expressed as:

$$\widehat{Z}(\mathbf{s}_0, t_0) = m_Z(\mathbf{s}_0, t_0) + \widehat{Z}'(\mathbf{s}_0, t_0), \tag{2.34}$$

where $m_Z(\mathbf{s}_0, t_0)$ is the estimated trend function and $\widehat{Z}'(\mathbf{s}_0, t_0)$ is the interpolated residual by means of STOK (Hengl, 2007).

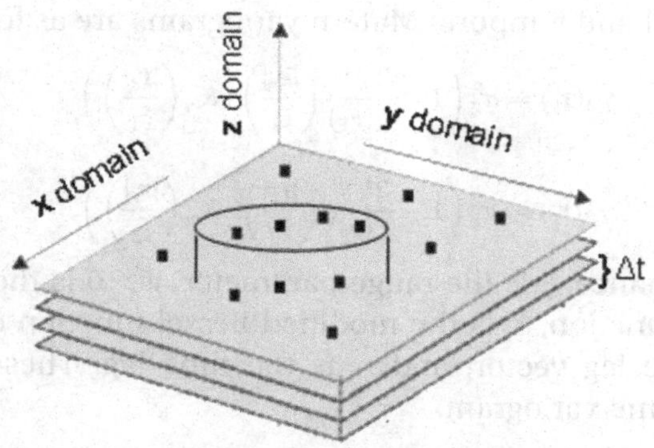

FIGURE 2.1 Representation of the space–time domain and of the space–time search neighborhood (after (Hengl, 2007)).

3. Spatiotemporal Prediction of Groundwater Level Data in Mires Basin, Crete, Greece

Mires basin of the Messara valley is a sparsely monitored basin. Since 1981 when a rapid increase of drip irrigation and increased pumping were started, only 10 wells were consistently monitored biannually. The basin is consistently overexploited and the result is a great drawdown of the water table; more than 35 m since 1981. Water resources in the area and especially groundwater are becoming depleted. Therefore the area is in need of spatial and temporal groundwater analyses.

Using spatiotemporal geostatistics the groundwater level dataset can be usefully exploited to identify the spatiotemporal behavior of the aquifer and to take useful information regarding the space–time data correlations for more accurate space–time predictions. Space–time geostatistical analysis employs the following steps: (1) space–time variogram calculation, (2) application of space–time kriging, STOK, for prediction, and (3) estimation of prediction accuracy.

First, the experimental variogram is determined. Then it is modeled with theoretical spatiotemporal variogram functions. The product sum model was herein applied as it has been successfully performed in other space–time studies providing better results than the other counterparts (De Iaco, 2010; Gething et al., 2007). Its main characteristic is its flexibility in modeling and estimation. The Matérn variogram model is chosen to simulate the spatial and temporal continuity of the data within the product–sum space–time model. The purely spatial geostatistical analysis of groundwater level data in previous works at different time periods have shown that the Matérn model describes very well the spatial correlation of the observed data (Varouchakis and Hristopulos, 2013).

The separable spatial and temporal Matérn variograms are as follows (Matérn, 1960):

$$\gamma_Z(\mathbf{r}_s) = \sigma_Z^2\left(1 - \frac{2^{1-\nu_1}}{\Gamma(\nu_1)}\left(\frac{|\mathbf{r}_s|}{\xi_1}\right)^{\nu_1} K_{\nu_1}\left(\frac{|\mathbf{r}_s|}{\xi_1}\right)\right), \tag{2.35}$$

$$\gamma_Z(r_t) = \sigma_Z^2\left(1 - \frac{2^{1-\nu_2}}{\Gamma(\nu_2)}\left(\frac{|r_t|}{\xi_2}\right)^{\nu_2} K_{\nu_2}\left(\frac{|r_t|}{\xi_2}\right)\right), \tag{2.36}$$

where $\sigma_Z^2 > 0$ is the variance, ξ is the range parameter, $\nu > 0$ is the smoothness parameter, Γ is the gamma function, K is the modified Bessel function of the second kind of order ν, $\mathbf{r}_s$ is the space lag vector, and r_t is the time lag. These are inserted in the product–sum space–time variogram.

4. Results and Discussion

Spatiotemporal geostatistical analysis of Mires basin groundwater level data has been applied to identify the spatiotemporal behavior of the aquifer since 1981 and to undertake predictions based on the space–time data correlations. The space–time experimental variogram is determined from the biannual groundwater level time series at the 10 sampling stations for the period 1981–2014. Validation of the estimates is performed for the wet period in the year 2015.

Theoretical space–time variogram model fitting on the experimental space–time variogram obtained from the observed data is presented in Fig. 2.2. The respective parameters for the variogram type are $\sigma_z^2 = 44.22$, $\xi_1 = 0.25$ ($\approx$3 km), $\nu_1 = 1.51$, $\xi_2 = 0.41$ ($\approx$12 months), $\nu_2 = 0.81$ with $k_1 = 0.45$, $k_2 = 1.5$, $k_3 = 1.25$.

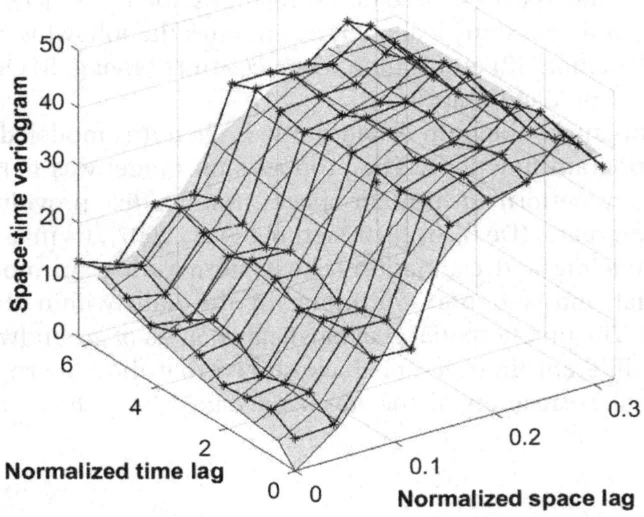

FIGURE 2.2 Space–time product–sum variogram fit using the Matérn structure.

The prediction involves STOK application to estimate the groundwater level at the specified locations and time during the wet period in the year 2015. The validation results in terms of absolute error estimation are presented in Table 2.1.

As it is presented in Table 2.1 STOK provides very good agreement with the reported values. The aquifer level map is then derived using STOK with the product–sum spatiotemporal variogram structure for the wet period in the year 2015, the last period of available data and the most recent to date. The contour map of groundwater level spatial variability in physical space is shown in Fig. 2.3. The map is constructed using estimates only at points inside the convex hull of the measurement locations.

Table 2.1 Absolute Error of Space–Time Ordinary Kriging Estimates for the Wet Period in the Year 2015

Well No.	Absolute Error (m)
G1	1.2
G2	1.02
G3	0.72
G4	1.21
G5	1.32
G6	1.32
G7	0.84
G8	0.95
G9	0.68
G10	0.73

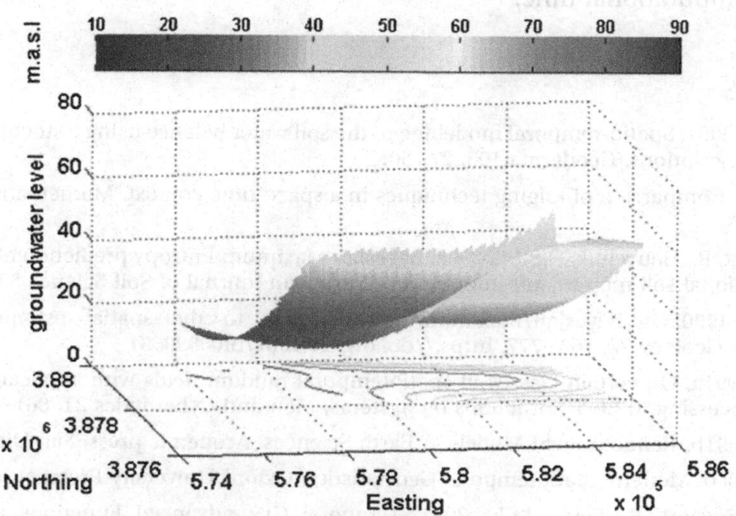

FIGURE 2.3 Map of estimated groundwater level (meters above sea level—masl) in the Mires basin using space–time ordinary kriging for the wet period in the year 2015.

The scope of this work was to model spatiotemporally Mires basin aquifer response since 1981. The model delivers an excellent variogram fit and very accurate estimates. The spatial correlation length is determined after variogram fitting equal to almost 3 km, and temporal length equal to almost 12 months. The latter denotes that the spatiotemporal model considers for the prediction process so the measurements of the wet as of the dry hydrological period. Fig. 2.3 presents the spatial variability of the groundwater level estimated based on the space–time correlations of the data that consider dynamic aquifer behavior.

5. Conclusions

Reliable space–time estimates are important for groundwater resources management. This work presented the space–time geostatistical analysis framework and examined the spatiotemporal modeling of groundwater level in a hydrological basin where the groundwater resources had been significantly depleted over the past 35 years. The spatiotemporal approach involved the application of the product–sum variogram function that has been successfully applied in other topics. This nonseparable spatiotemporal structure fits very well the experimental space–time variogram of the groundwater level capturing the space–time correlations of the available data. The STOK estimates presented accurately the groundwater level variability for the examined validation period and provided the spatial distribution of the aquifer level at ungauged locations for the wet period in the year 2015. The examined approach was shown to provide a reliable alternative in spatiotemporal modeling of aquifer level. Another advantage is that it requires fewer data than a numerical model to represent the head field in less computational time.

References

Bierkens, M.F.P., 2001. Spatio-temporal modelling of the soil water balance using a stochastic model and soil profile descriptions. Geoderma 103, 27–50.

Bogaert, P., 1996. Comparison of kriging techniques in a space-time context. Mathematical Geology 28, 73–86.

Brus, D.J., Bogaert, P., Heuvelink, G.B.M., 2008. Bayesian Maximum Entropy prediction of soil categories using a traditional soil map as soft information. European Journal of Soil Science 59, 166–177.

Christakos, G., 1990. A Bayesian/maximum-entropy view to the spatial estimation problem. Mathematical Geology 22, 763–777. https://doi.org/10.1007/bf00890661.

Christakos, G., 1991a. On certain classes of spatiotemporal random fields with applications to space - time data processing. IEEE Transactions on Systems, Man and Cybernetics 21, 861–875.

Christakos, G., 1991b. Random Field Models in Earth Sciences. Academic press, San Diego.

Christakos, G., 2000. Modern Spatiotemporal Geostatistics. Oxford University Press, New York.

Christakos, G., Bogaert, P., Serre, M.L., 2001. Temporal GIS: advanced Functions for Field-Based Applications, vol. 1. Springer Verlag, Berlin.

Christakos, G., Hristopulos, D.T., 1998. Spatiotemporal Environmental Health Modelling: A Tractatus Stochasticus. Kluwer, Boston.

Christakos, G., Serre, M.L., 2000. BME analysis of spatiotemporal particulate matter distributions in North Carolina. Atmospheric Environment 34, 3393–3406. https://doi.org/10.1016/s1352-2310(00)00080-7.

Cressie, N., 1993. Statistics for Spatial Data, revised ed. Wiley, New York.

Cressie, N., Huang, H.C., 1999. Classes of nonseparable, spatio-temporal stationary covariance functions. Journal of the American Statistical Association 94, 1330–1340.

De Cesare, L.D., Myers, D.E., Posa, D., 2001. Estimating and modeling space-time correlation structures. Statistics & Probability Letters 51, 9–14.

De Cesare, L.D., Myers, D.E., Posa, D., 2002. FORTRAN programs for space-time modeling. Computers & Geosciences 28, 205–212.

De Iaco, S., 2010. Space-time correlation analysis: a comparative study. Journal of Applied Statistics 37, 1027–1041.

De Iaco, S., Myers, D.E., Posa, D., 2001. Space-time analysis using a general product-sum model. Statistics & Probability Letters 52, 21–28.

De Iaco, S., Myers, D.E., Posa, D., 2002a. Nonseparable space-time covariance models: some parametric families. Mathematical Geology 34, 23–42.

De Iaco, S., Myers, D.E., Posa, D., 2002b. Space-time variograms and a functional form for total air pollution measurements. Computational Statistics & Data Analysis 41, 311–328.

Dimitrakopoulos, R., Luo, X., 1994. Spatiotemporal modeling: covariances and ordinary kriging systems. In: Dimitrakopoulos, R. (Ed.), Geostatistics for the Next Century. Kluwer, Dordrecht, pp. 88–93.

Fischer, M.M., Getis, A., 2010. Handbook of Applied Spatial Analysis: Software Tools, Methods and Applications. Springer Verlag, Berlin.

Gething, P.W., Atkinson, P.M., Noor, A.M., Gikandi, P.W., Hay, S.I., Nixon, M.S., 2007. A local space-time kriging approach applied to a national outpatient malaria data set. Computers & Geosciences 33, 1337–1350.

Giraldo Henao, R., 2009. Geostatistical Analysis of Functional Data (PhD). Universitat Politechnica de Catalunya.

Gneiting, T., 2002. Nonseparable, stationary covariance functions for space-time data. Journal of the American Statistical Association 97, 590–600.

Gneiting, T., Genton, M.G., Guttorp, P., 2007. Geostatistical Space-time Models, Stationarity, Separability and Full Symmetry. Department of Statistics, University of Washington.

Goovaerts, P., 1997. Geostatistics for Natural Resources Evaluation. Oxford University Press, New York.

Hengl, T., 2007. A Practical Guide to Geostatistical Mapping of Environmental Variables. Office for Official Publications of the European Communities, Luxembourg.

Hengl, T., Heuvelink, G.B.M., Perčec Tadić, M., Pebesma, E.J., 2011. Spatio-temporal prediction of daily temperatures using time-series of MODIS LST images. Theoretical and Applied Climatology 107, 1–13.

Heuvelink, G.B.M., Egmond, F.M., 2010. Space–time geostatistics for precision agriculture: a case study of NDVI mapping for a Dutch potato field. In: Oliver, M.A. (Ed.), Geostatistical Applications for Precision Agriculture. Springer, Netherlands, pp. 117–137. https://doi.org/10.1007/978-90-481-9133-8_5.

Heuvelink, G.B.M., Griffith, D.A., 2010. Space-time geostatistics for geography: a case study of radiation monitoring across parts of Germany. Geographical Analysis 42, 161–179.

Hoogland, T., Heuvelink, G.B.M., Knotters, M., 2010. Mapping water-table depths over time to assess desiccation of groundwater-dependent ecosystems in The Netherlands. Wetlands 30, 137–147.

Hristopulos, D.T., 2002. New anisotropic covariance models and estimation of anisotropic parameters based on the covariance tensor identity. Stochastic Environmental Research and Risk Assessment 16, 43–62.

Jost, G., Heuvelink, G.B.M., Papritz, A., 2005. Analysing the space-time distribution of soil water storage of a forest ecosystem using spatio-temporal kriging. Geoderma 128, 258–273.

Kolovos, A., Angulo, J., Modis, K., Papantonopoulos, G., Wang, J.-F., Christakos, G., 2012. Model-driven development of covariances for spatiotemporal environmental health assessment. Environmental Monitoring and Assessment 1–17. https://doi.org/10.1007/s10661-012-2593-1.

Kolovos, A., Christakos, G., Hristopulos, D.T., Serre, M.L., 2004. Methods for generating non-separable spatiotemporal covariance models with potential environmental applications. Advances in Water Resources 27, 815–830.

Kyriakidis, P., Journel, A., 1999. Geostatistical space-time models: a review. Mathematical Geology 31, 651–684.

Kyriakidis, P.C., Journel, A.G., 2001a. Stochastic modeling of atmospheric pollution: a spatial time-series framework. Part I: methodology. Atmospheric Environment 35, 2331–2337. https://doi.org/10.1016/s1352-2310(00)00541-0.

Kyriakidis, P.C., Journel, A.G., 2001b. Stochastic modeling of atmospheric pollution: a spatial time-series framework. Part II: application to monitoring monthly sulfate deposition over Europe. Atmospheric Environment 35, 2339–2348. https://doi.org/10.1016/s1352-2310(00)00540-9.

Lee, S.-J., Wentz, E.A., 2008. Applying Bayesian maximum entropy to extrapolating local-scale water consumption in Maricopa county, Arizona. Water Resources Research 44, W01401. https://doi.org/10.1029/2007wr006101.

Lee, S.J., Wentz, E.A., Gober, P., 2010. Space-time forecasting using soft geostatistics: a case study in forecasting municipal water demand for Phoenix, Arizona. Stochastic Environmental Research and Risk Assessment 24, 283–295.

Matérn, B., 1960. Spatial variation. Meddelanden från Statens Skogsforskningsinstitut 49, 1–144.

Mendoza-Cazares, E.Y., Herrera-Zamarron, G.D., 2010. Spatiotemporal estimation of hydraulic head using a single spatiotemporal random function model. Tecnologia y Ciencias del Agua 1, 87–111.

Monin, A., Yaglom, A., 1975. Statistical Fluid Mechanics: Mechanics of Turbulence, vol. 2. MIT Press, Cambridge.

Myers, D.E., 2002. Space-time correlation models and contaminant plumes. Environmetrics 13, 535–553.

Myers, D.E., De Iaco, S., Posa, D., De Cesare, L., 2002. Space-time radial basis functions. Computers & Mathematics With Applications 43, 539–549.

Myers, D.E., Journel, A., 1990. Variograms with zonal anisotropies and noninvertible kriging systems. Mathematical Geology 22, 779–785.

Nunes, C., Soares, A., 2005. Geostatistical space-time simulation model for air quality prediction. Environmetrics 16, 393–404.

Park, M.S., Fuentes, M., 2008. Testing lack of symmetry in spatial-temporal processes. Journal of Statistical Planning and Inference 138, 2847–2866.

Porcu, E., Gregori, P., Mateu, J., 2006. Nonseparable stationary anisotropic space-time covariance functions. Stochastic Environmental Research and Risk Assessment 21, 113–122.

Porcu, E., Mateu, J., Saura, F., 2008. New classes of covariance and spectral density functions for spatio-temporal modelling. Stochastic Environmental Research and Risk Assessment 22, 65–79.

Rodriguez-Iturbe, I., Mejia, M.J., 1974. The design of rainfall networks in time and space. Water Resources Research 10, 713–728.

Rouhani, S., Hall, T.J., 1989. Space–time kriging of groundwater data. In: Armstrong, M. (Ed.), Geostatistics, vol. 2. Kluwer Academic Publishers, The Netherlands, pp. 639–651.

Rouhani, S., Myers, D., 1990. Problems in space-time kriging of geohydrological data. Mathematical Geology 22, 611–623.

Skøien, J.O., Blöschl, G., 2007. Spatiotemporal topological kriging of runoff time series. Water Resources Research 43.

Snepvangers, J.J.J.C., Heuvelink, G.B.M., Huisman, J.A., 2003. Soil water content interpolation using spatio-temporal kriging with external drift. Geoderma 112, 253–271.

Stein, A., 1998. Analysis of space-time variability in agriculture and the environment with geostatistics. Statistica Neerlandica 52, 18–41.

Varouchakis, E.A., 2017. Spatiotemporal geostatistical modelling of groundwater level variations at basin scale: a case study at Crete's Mires Basin. Hydrology Research. https://doi.org/10.2166/nh.2017.146.2017.2146.

Varouchakis, E.A., Hristopulos, D.T., 2013. Improvement of groundwater level prediction in sparsely gauged basins using physical laws and local geographic features as auxiliary variables. Advances in Water Resources 52, 34–49.

Varouchakis, E.A., Hristopulos, D.T., 2017. Comparison of spatiotemporal variogram functions based on a sparse dataset of groundwater level variations. Spatial Statistics. https://doi.org/10.1016/j.spasta.2017.1007.1003.

Vyas, V.M., Christakos, G., 1997. Spatiotemporal analysis and mapping of sulfate deposition data over Eastern U.S.A. Atmospheric Environment 31, 3623–3633. https://doi.org/10.1016/s1352-2310(97)00172-6.

Yu, H.L., Chen, J.C., Christakos, G., Jerrett, M., 2009. BME estimation of residential exposure to ambient PM10 and ozone at multiple time scales. Environmental Health Perspectives 117, 537–544.

Yu, H.L., Yang, S.J., Yen, H.J., Christakos, G., 2011. A spatio-temporal climate-based model of early dengue fever warning in southern Taiwan. Stochastic Environmental Research and Risk Assessment 25, 485–494.

3

Large-Scale Exploratory Analysis of the Spatiotemporal Distribution of Climate Projections: Applying the STRIVIng Toolbox

Vitali Diaz[1,2], Gerald Corzo[1], José R. Pérez[3]

[1]UNESCO-IHE INSTITUTE FOR WATER EDUCATION, DELFT, THE NETHERLANDS; [2]WATER RESOURCES SECTION, DELFT UNIVERSITY OF TECHNOLOGY, DELFT, THE NETHERLANDS; [3]INSTITUTO NACIONAL DE RECURSOS HIDRÁULICOS (INDRHI), SANTO DOMINGO, DOMINICAN REPUBLIC

1. Introduction

Precipitation and temperature projections over the next few decades indicate that on earth their spatiotemporal distribution will undergo changes (IPCC, 2014; Moss et al., 2010; Najafi and Moradkhani, 2015; Taylor et al., 2012). These variations might modify the way in which extreme hydrological events (EHEs) occur, affecting their frequency and intensity (e.g., Coumou and Rahmstorf, 2012; Trenberth, 2012). Because EHEs, such as droughts and floods, have a great negative impact on human activities, a better understanding of the expected spatiotemporal variability of precipitation and temperature is necessary.

Relatively recent studies have been carried out to analyze the spatiotemporal changes of future precipitation and temperature considering the entire planet or a number of continents (e.g., Boer, 2009; Hawkins and Sutton, 2009; Milly et al., 2005). There are also applications at lower scales (e.g., Wang et al., 2016). These data have been considered a challenge for big data in many studies. This relates to the amount of information that can be handled and the processes involved in the analysis.

In this study we present an analysis using the Spatio-TempoRal distribution and Interannual VarIability of projections (STRIVIng) toolbox for statistical exploratory analysis of future precipitation and temperature. The toolbox provides a set of elements for numerical and visual comparison of the baseline and projections. STRIVIng is designed to work with monthly values and various spatial resolutions. In this document, large-scale applications are presented following a standard step-by-step exploratory analysis.

Spatiotemporal Analysis of Extreme Hydrological Events. https://doi.org/10.1016/B978-0-12-811689-0.00003-3

To illustrate the use of STRIVIng, three case studies were undertaken: Dominican Republic, Mexico, and Amazon basin, to try to cover regions of different sizes. Following this introductory section, concepts and the description of the toolbox are shown. Subsequently, data, case studies, and the application are described. Finally, results and conclusions are presented.

2. Framework (Global Climate Model Projections)

A type of model to simulate climate is the so-called General Circulation Model (GCMs), which makes use of mathematical formulations to reproduce the general circulation of the atmosphere. To analyze the possible implications of the human and natural influences over climate, GCMs are forced, in a systematic way, to analyze possible futures under the scenarios approach (Moss et al., 2010). In this way, climate projections are estimated for each scenario. The design of scenarios and GCM setups are coordinated worldwide by the Coupled Model Intercomparison Project (CMIP), where several modeling teams participate (Meehl et al., 2014).

Projection data are available for each GCM and each scenario. Due to the information of these variables is usually on a coarse spatial and temporal scale, in addition to presenting a deviation in value compared to ground observations, at least two steps are necessary before use: downscaling and bias correction. Downscaling refers to the procedure of changing the coarse scale to a finer one. Bias correction is the procedure of tuning simulated variables in the atmosphere with those gathered on the ground.

There are a couple of examples of research projects where the adjustment (i.e., downscaling and calibration) of *P* and *T* projections was within the scope: ENSEMBLES (ensembles-eu.metoffice.com) and Regional Climate, Water, Energy Resources and uncertainties (RIWER 2030 project; www.lthe.fr/RIWER 2030), where the target was Europe. Global research projects included the Integrated Project Water and Global Change (WATCH, www.eu-watch.org) and WorldClim (worldclim.org). From all these projects, data of adjusted *P* and *T* are freely available for download or on request.

As Meehl et al. (2014) point out "to better understand the past, present, and future climate, the state-of-the-art climate model simulations are compared to gain insights into the processes, mechanisms, and consequences of climate variability." One way to carry out this comparative analysis is through the examination of the statistics of adjusted climate variables, as well as the investigation of their expected spatial distribution. The toolbox introduced in this document helps to carry out comparative analysis. A description of the toolbox is given next.

3. STRIVIng Methodology

In general, the methodology involves two steps: processing and visualization (Fig. 3.1). In the first part, input data are processed to calculate the statistics and average values of

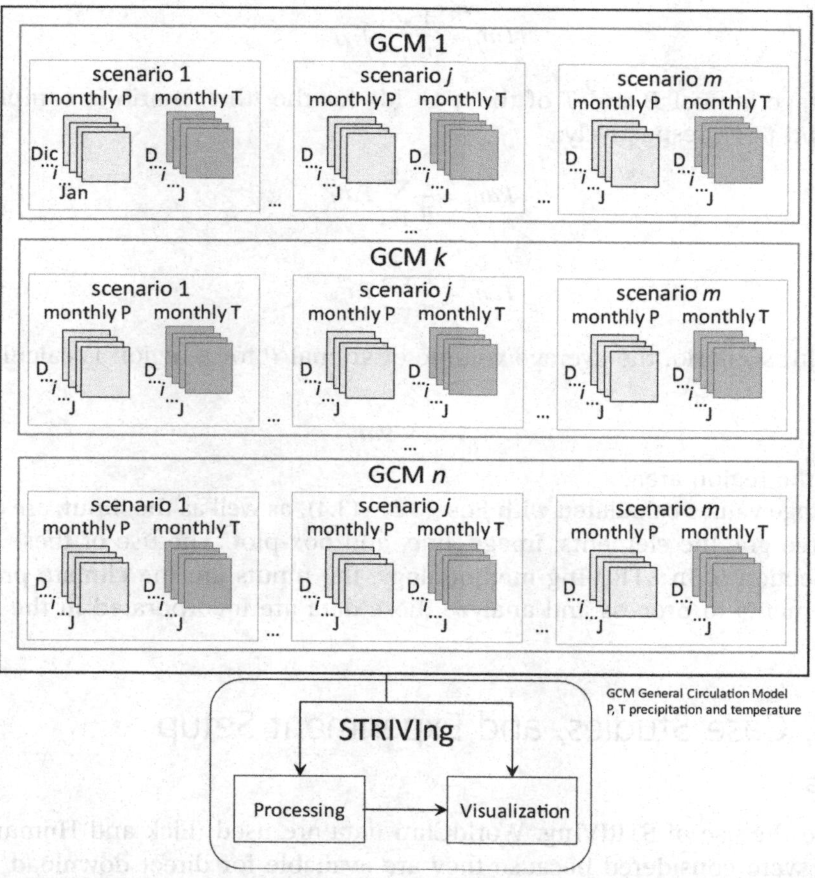

FIGURE 3.1 Schematic overview of the STRIVIng toolbox. *GCM*, General Circulation Model; *P*, precipitation; *T*, temperature.

adjusted projections for a study area. The second part consists of a set of graphic elements for the visualization of spatial distribution, as well as monthly average values of precipitation and temperature for the study area.

Regarding processing, consider that there are n GCMs (Fig. 3.1). Each kth GCM has the information of m scenarios. Each scenario, in turn, has data of the long-term mean precipitation and mean temperature by month from January to December (J to D). Also for each scenario, annual precipitation and temperature are available for reference and future periods. The long-term mean precipitation (Pm) and mean temperature (Tm) for the ith month and the jth scenario, also called ensemble mean, are calculated with Eqs. (3.1) and (3.2), respectively:

$$Pm_{i,j} = \frac{1}{n}\sum_{k=1}^{n} P_{i,j,k} \tag{3.1}$$

$$Tm_{i,j} = \frac{1}{n}\sum_{k=1}^{n} T_{i,j,k}. \tag{3.2}$$

The average annual P and T of the n GCMs for the jth scenario is computed with Eqs. (3.3) and (3.4), respectively:

$$Pan_j = \frac{1}{n}\sum_{k=1}^{n} Pa_{j,k} \tag{3.3}$$

$$Tan_j = \frac{1}{n}\sum_{k=1}^{n} Ta_{j,k}. \tag{3.4}$$

For the jth scenario, the average volume of annual P for a region is calculated with Eq. (3.5):

$$V_j = A \times Pan_j \tag{3.5}$$

where A is the region area.

The average values calculated with Eqs. (3.1)–(3.4), as well as the input, are examined through three graphic elements: image, line, and box-plot. The use of these graphs is shown in Section 5. In STRIVIng methodology, the inputs are the climate projections, and the elements to process and analyze these data are incorporated in the STRIVIng toolbox.

4. Data, Case Studies, and Experiment Setup

4.1 Data

To illustrate the use of STRIVIng, WorldClim data are used (Fick and Hijmans, 2017). These data were considered because they are available for direct download and their spatial resolution is lower than half a degree. A detailed description of data and the adjustment procedure can be found in Fick and Hijmans (2017) and at www.worldclim.org/downscaling.

WorldClim projections correspond to the fifth assessment of CMIP (IPCC, 2014). In CMIP5, four Representative Concentration Pathways (RCPs) of greenhouse gas (GHG) concentration trajectories are taken into account for climate modeling. RCPs refer to four likely climate futures where diverse amounts of GHG are emitted at different times. As Meinshausen et al. (2011) highlight "RCP 2.6 assumes that global annual GHG emissions peak between 2010-20, with emissions declining substantially thereafter. Emissions in RCP 4.5 peak around 2040, then decline. In RCP 6, emissions peak around 2080, then decline. In RCP 8.5, emissions continue to rise throughout the 21st century."

Table 3.1 WorldClim Data (Fick and Hijmans, 2017)

GCM	RCP 2.6	RCP 4.5	RCP 6.0	RCP 8.5
$^{+}$ACCESS1-0		*		*
BCC-CSM1-1	*	*	*	*
CCSM4	*	*	*	*
CESM1-CAM5-1-FV2		*		
$^{+}$CNRM-CM5	*	*		*
GFDL-ESM2G	*	*	*	
GFDL-CM3	*	*		*
GISS-E2-R	*	*	*	*
HadGEM2-AO	*	*	*	*
HadGEM2-ES	*	*	*	*
HadGEM2-CC		*		*
INMCM4		*		*
IPSL-CM5A-LR	*	*	*	*
$^{+}$MIROC5	*	*	*	*
MRI-CGCM3	*	*	*	*
$^{+}$MIROC-ESM-CHEM	*	*	*	*
MPI-ESM-LR	*	*		*
$^{+}$MIROC-ESM	*	*	*	*
NorESM1-M	*	*	*	*

*, indicates the availability of data. +, for noncommercial use. GCM and RCP stand for General Circulation Model and Representative Concentration Pathway, respectively.

Future *P* and *T* of WorldClim (Table 3.1) are available for 19 GCMs, four RCPs (scenarios), and the aggregate periods of 2050 (2041–60) and 2070 (2061–80). Data are arranged in four spatial resolutions: 30 s ($\sim$1 km^2), 2.5 s ($\sim$20 km^2), 5 s ($\sim$80 km^2), and 10 min ($\sim$340 km^2 at the equator). The baseline information corresponds to the period 1950–2000. In this chapter, 10-min data of both baseline and projections were used for the large-scale applications presented hereafter.

4.2 Case Studies

For the application, two countries and one large basin were considered: Dominican Republic, Mexico, and Amazon basin (Fig. 3.2). The Amazon basin is located in the territories of Brazil, Bolivia, Peru, Ecuador, Colombia, Venezuela, Guyana, Suriname, and French Guiana. The Amazon River and its tributaries drain throughout this basin into the Atlantic Ocean. In Table 3.2, the main characteristics of the cases studies are presented.

4.3 Experiment Setup

P and *T* projections of the study areas were processed taking into account the 19 GCMs and four RCPs of WorldClim data, as well as the information of the baseline period of 1970–2000. In Section 6, some relevant findings are presented. To extract and process

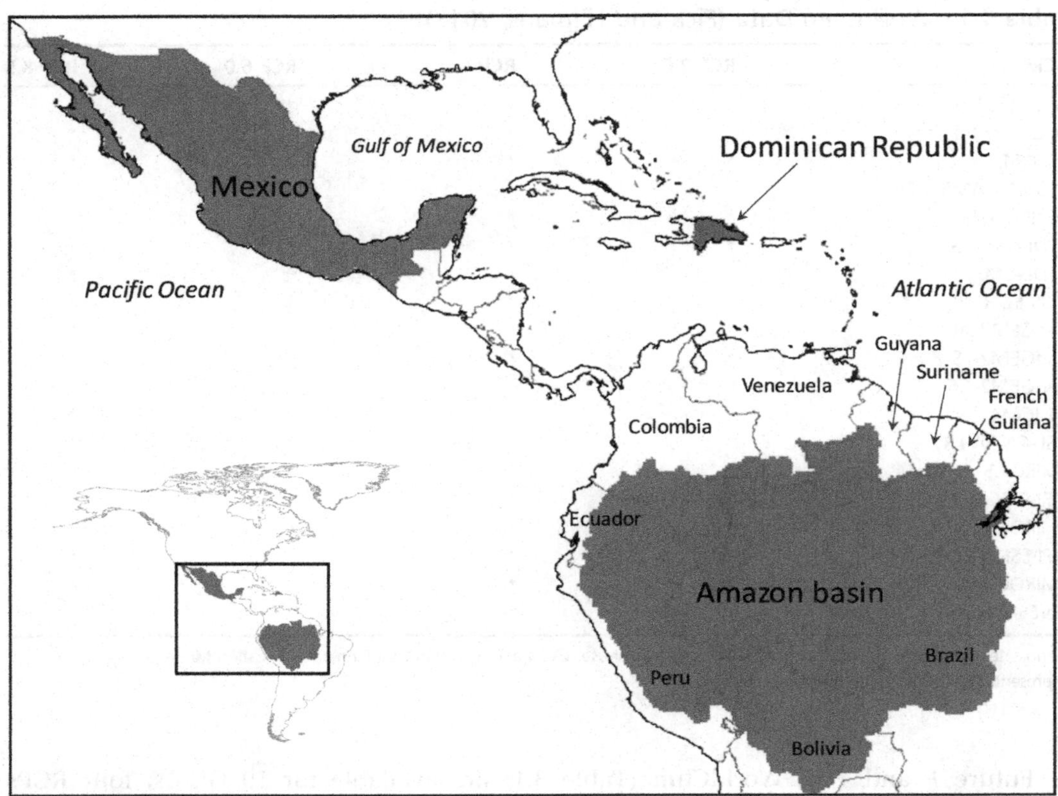

FIGURE 3.2 Mexico, Dominican Republic, and Amazon basin location.

Table 3.2 Characteristics of Case study Areas

Case Study	Area (km^2)	Mean Annual Precipitation (mm)	Mean Annual Temperature (°C)
Dominican Republic[1]	48,315	1500	25
Mexico[2]	1,972,550	780	21
Amazon basin[3]	6,150,000	2500	24–26

Sources: (1) US Library of Congress; (2) National Meteorological Service (SMN, Abbreviation in Spanish), Mexico; (3) Food and Agriculture Organization of the United Nations (FAO).

the data, three masks were built with the same spatial resolution as the data, i.e., 10 min. Long-term mean *P* and *T* from January to December (J to D) were computed with Eqs. (3.1) and (3.2). The mean annual *P* and *T* were calculated by using Eqs. (3.3) and (3.4), respectively. To estimate the annual water volume with Eq. (3.5), the next areas where considered: Dominican Republic (48,670 km^2), Mexico (1,972,550 km^2), and Amazon basin (6,171,148.7 km^2). These areas are slightly different from those reported in the sources (Table 3.2), but correspond to the shape files used for the construction of the masks. Water volumes are shown in billions of cubic meters (10^9 m^3).

5. Results and Discussion

5.1 Dominican Republic

For both periods 2050 (2041–60) and 2070 (2061–80), most GCMs agree that the greatest changes in the spatial distribution of P will occur on the north coast. While on the border with Haiti, the condition practically remains stable, Figs. 3.3 and 3.4, respectively. For the 2050 period, the northern coast will experience an increase in annual rainfall, while a decrease is expected on the southern coast. On the other hand, for the period of 2070, it is expected that on the north coast the magnitude of the annual rains will be of the same order as the historical ones. The south coast will experience a decrease with respect to those of the baseline period.

Fig. 3.5 shows that the long-term mean P for the 12 months, in general, will show a decrease in the coming decades, mainly in the months of May to October. June and July are the months that present the greatest fall. The May rainfall projection shows the greatest disagreement among GCMs.

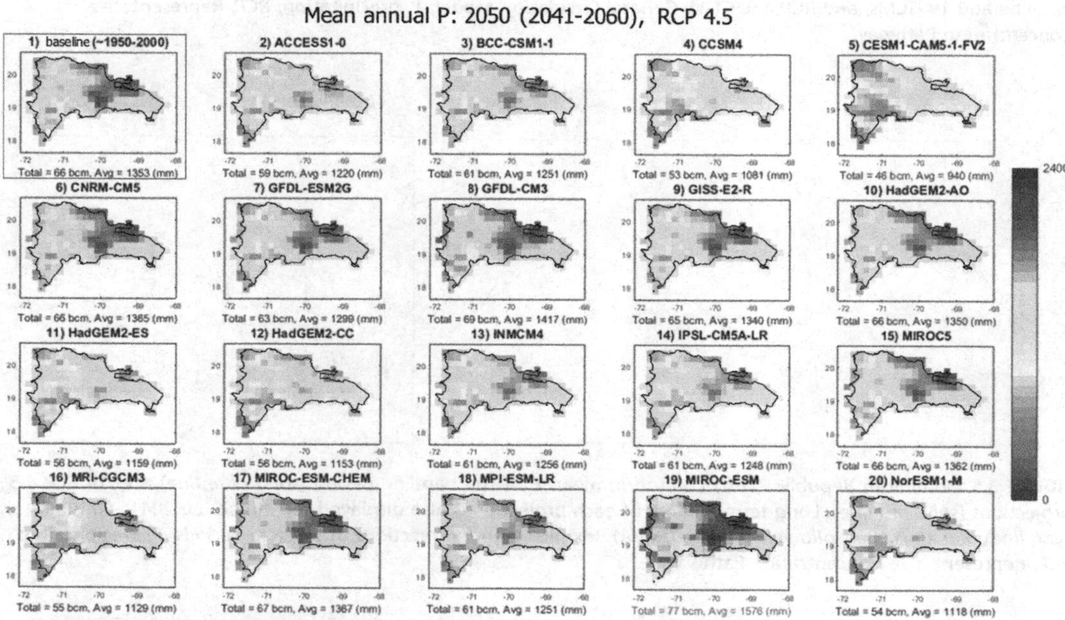

FIGURE 3.3 Spatial distribution of mean annual P (mm) over Dominican Republic for the period 2041–60 (2050): baseline and 19 GCMs, and RCP 4.5. *GCM*, General Circulation Model; *P*, precipitation; *RCP*, Representative Concentration Pathway.

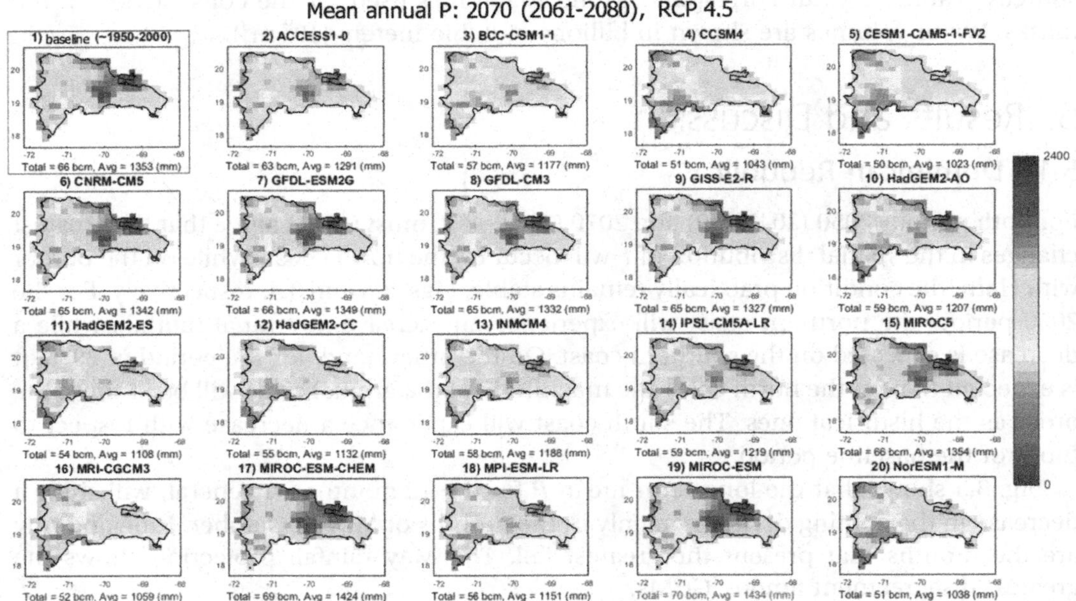

FIGURE 3.4 Spatial distribution of mean annual *P* (mm) over Dominican Republic for the period 2061–80 (2070): baseline and 19 GCMs, and RCP 4.5. *GCM*, General Circulation Model; *P*, precipitation; *RCP*, Representative Concentration Pathway.

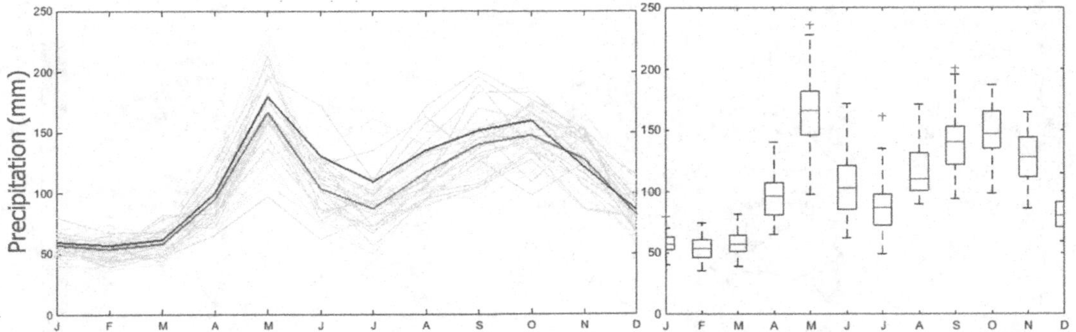

FIGURE 3.5 Dominican Republic. (*Left*) Long-term mean *P* of the baseline period (*solid blue line*) and 19 RCP 4.5 projections (*solid red line*). Long-term mean *P* of each projection is also displayed for the period 2041–60 (*solid cyan line*) and 2061–80 (*solid yellow line*). (*Right*) Box-plot with *P* projections of the two periods. *P*, Precipitation; *RCP*, Representative Concentration Pathway.

Regarding temperatures, Figs. 3.6 and 3.7 indicate that the country will suffer an increase in the mean annual values in practically the whole territory for the periods 2041–60 and 2061–80, respectively. Only in the central part will the condition remain more stable in both cases. Fig. 3.8 shows that the long-term mean T will increase to around 2 degrees in all 12 months.

5.2 Mexico

The spatial distribution of the mean annual P agrees with most of the GCMs for the period 2041–60 (Fig. 3.9). The maximum magnitudes of rainfall are observed on the south coast with the Gulf of Mexico and on the southwest coast with the Pacific Ocean. The lowest values are observed in the north and northwest of the country, on the border with the United States. The area with most disagreement is the southwest coast with the Pacific Ocean.

According to the four RCPs (Fig. 3.10), the long-term mean P of projections of the 12 months will remain with similar magnitude throughout the year, except for the months of June to September (J to S), where the models show mainly a drop in the values. July is the month that shows the greatest fall; this drop increases from RCP 2.6 to

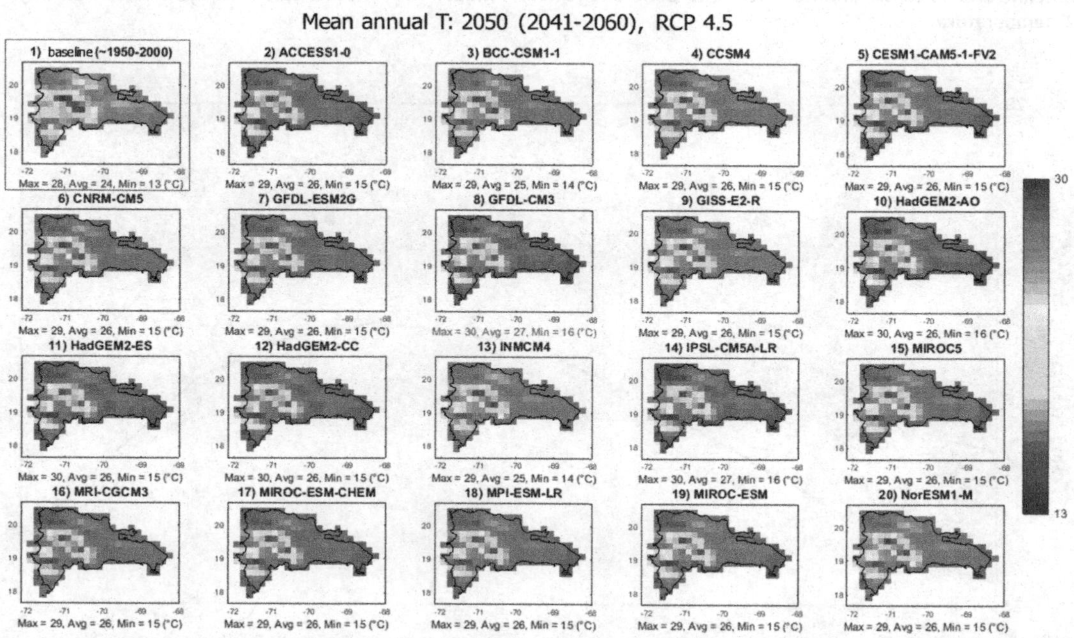

FIGURE 3.6 Spatial distribution of mean annual T (°C) over Dominican Republic for the period 2041–60 (2050): baseline and 19 GCMs, and RCP 4.5. *GCM*, General Circulation Model; *RCP*, Representative Concentration Pathway; *T*, temperature.

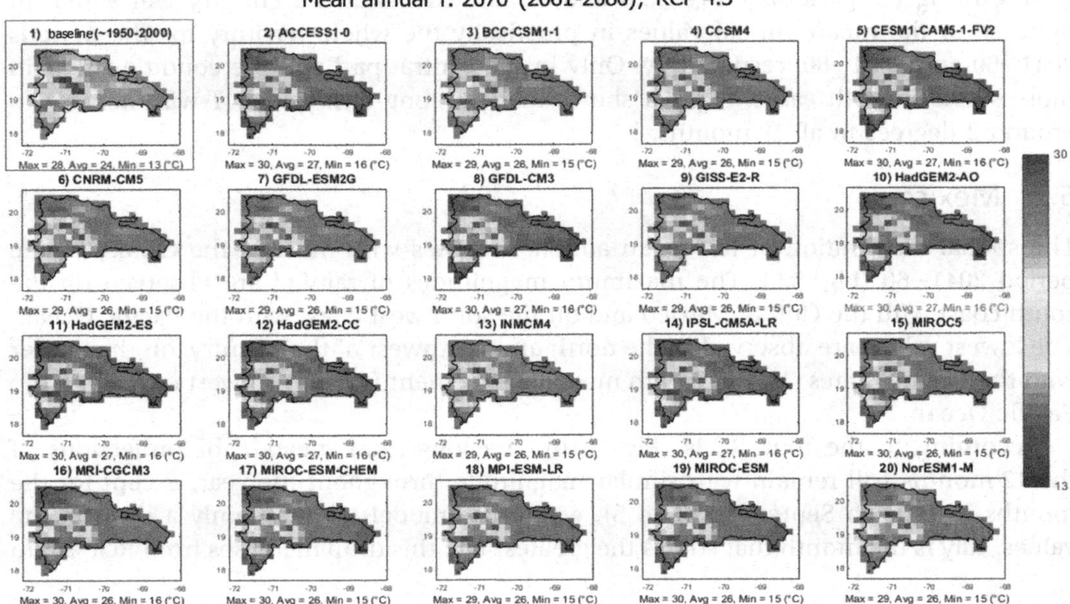

FIGURE 3.7 Spatial distribution of mean annual *T* (°C) over Dominican Republic for the period 2061–80 (2070): baseline and 19 GCMs, and RCP 4.5. *GCM*, General Circulation Model; *RCP*, Representative Concentration Pathway; *T*, temperature.

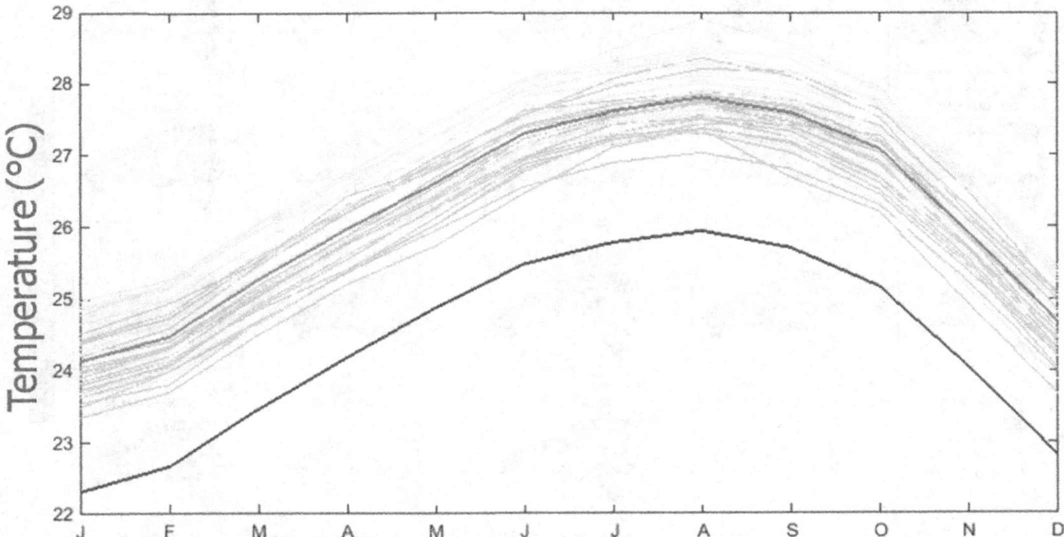

FIGURE 3.8 Dominican Republic: long-term mean *T* of the baseline period (*solid blue line*) and 19 RCP 4.5 projections (*solid red line*). Long-term mean *T* of each projection is also displayed for the period 2041–60 (*solid cyan lines*) and 2061–80 (*solid yellow lines*). *RCP*, Representative Concentration Pathway; *T*, temperature.

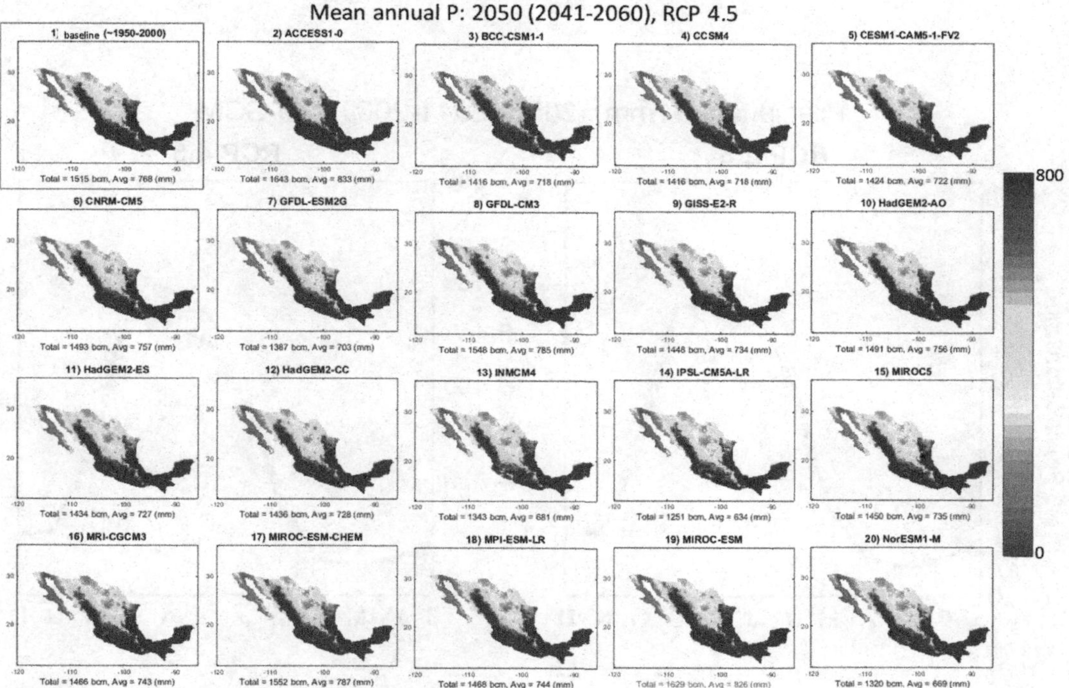

FIGURE 3.9 Spatial distribution of mean annual *P* (mm) over Mexico for the period 2041–60 (2050): baseline and 19 GCMs, and RCP 4.5. *GCM*, General Circulation Model; *P*, precipitation; *RCP*, Representative Concentration Pathway.

8.5. On the other hand, September shows an increase, whose value seems to be lower from route 2.6 to 8.5. July and September are the months with the poorest agreement between GCMs.

Concerning mean annual *T*, Fig. 3.11 shows that most GCMs indicate a rise in the temperatures throughout the country for the period 1941–2060. On the coasts and in the north, the greatest values are observed. Fig. 3.12 also points out an increment in the temporal distribution; values of long-term mean *T* of projections from January to December are larger than those of the baseline period. From June to September, the main increases are observed.

5.3 Amazon Basin

According to the baseline period, the spatial distribution of mean annual *P* shows different values along the basin. The largest appear to be in the northwest and center, followed by the rains in the northwest. The lowest are in the southwest part. According to RCP 4.5, most GCMs indicate a change in the spatial distribution of annual rainfall for

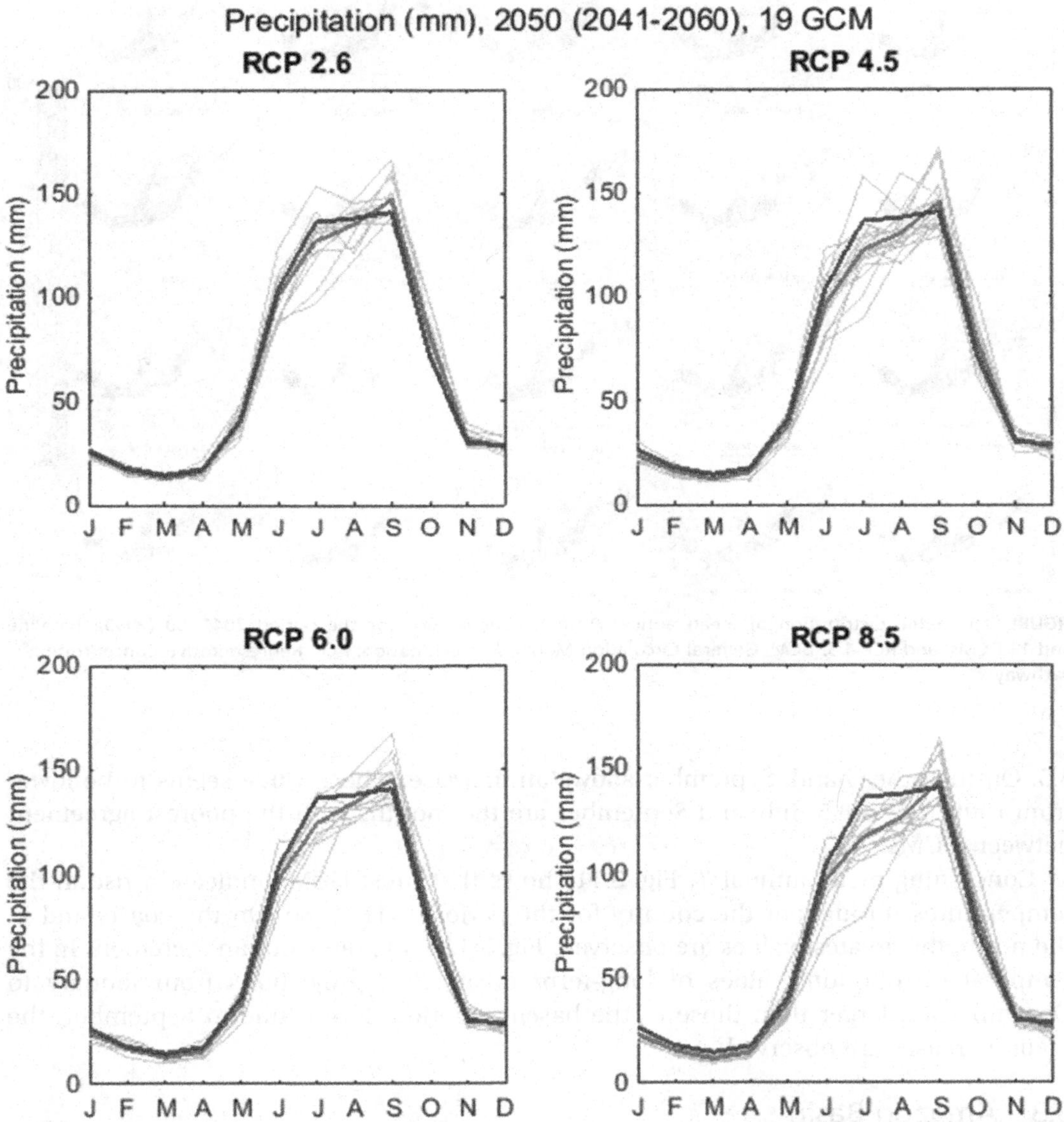

FIGURE 3.10 Mexico: long-term mean *P* of the baseline period (*solid blue line*) and the 19 RCP 2.6, 4.5, 6, and 8.5 projections (*solid red line*). Long-term mean *P* of each projection is also displayed for the period 2041–60 (*solid gray lines*). *P*, Precipitation; *RCP*, Representative Concentration Pathway.

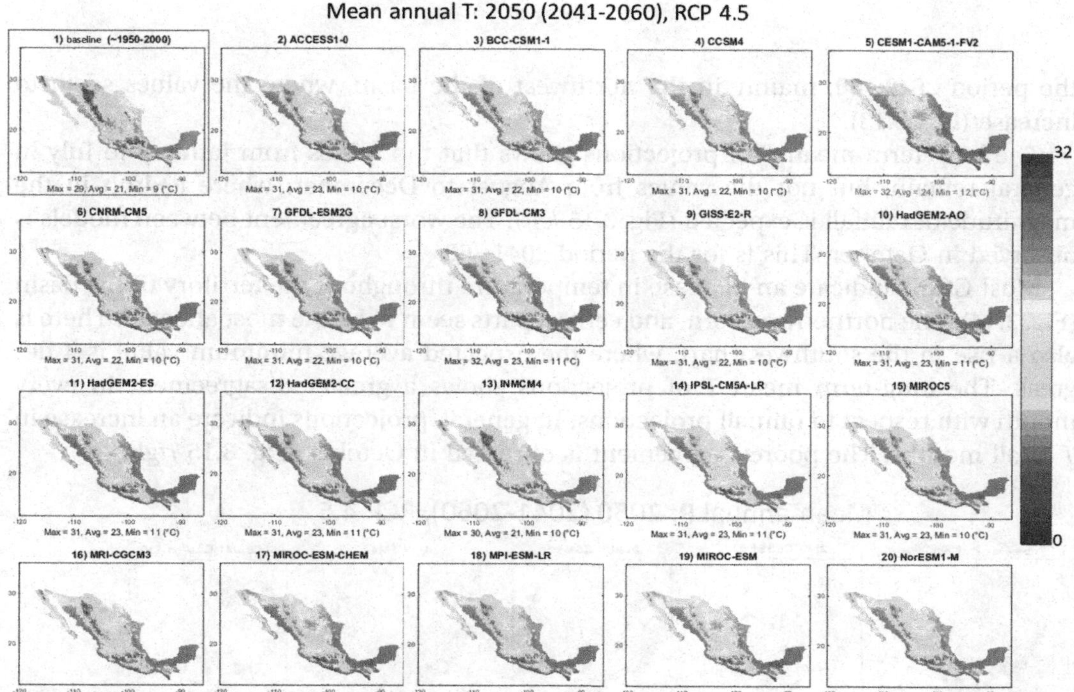

FIGURE 3.11 Spatial distribution of mean annual *T* (°C) over Mexico for the period 2041–60 (2050): baseline and 19 GCMs, and RCP 4.5. *GCM*, General Circulation Model; *RCP*, Representative Concentration Pathway; *T*, temperature.

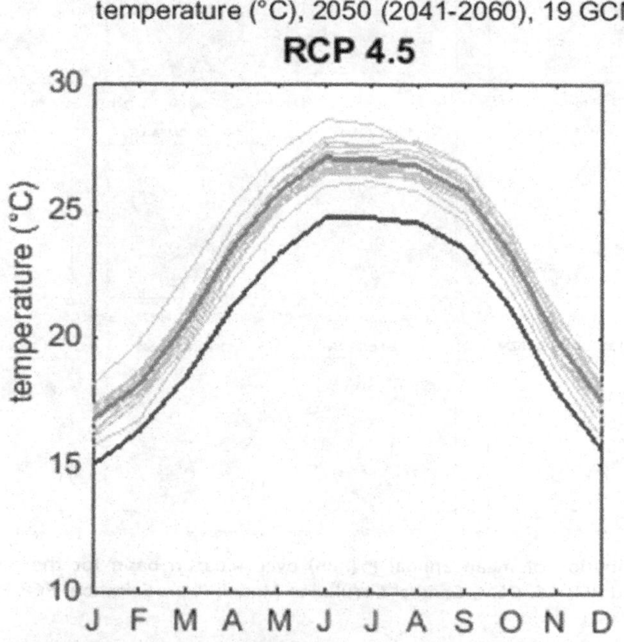

FIGURE 3.12 Mexico: long-term mean *T* of the baseline period (*solid blue line*) and the 19 RCP 4.5 projections (*solid red line*). Long-term mean *P* of each projection is also displayed for the period 2041–60 (*solid gray lines*). *P*, Precipitation; *RCP*, Representative Concentration Pathway; *T*, temperature.

the period 2041–60, mainly in the northwest of the basin, where the values seem to increase (Fig. 3.13).

The long-term mean *P* of projections shows that the values from January to July in general remain, but not the values from August to December, where a drop in the magnitude of rainfall is expected (Fig. 3.15 *left*). The worst agreement between models is observed in October. This is for the period 2041–60.

Most GCMs indicate an increase in temperature throughout the territory of the basin (Fig. 3.14). The northern, eastern, and central parts seem to be the most affected. There is also a rise in the southwest part, where the expected average minimum value is 4 degrees. The long-term mean *T* of projections shows a greater disagreement in every month with respect to rainfall projections. In general, projections indicate an increase in *T* in all months. The poorest agreement is observed in October (Fig. 3.15 *right*).

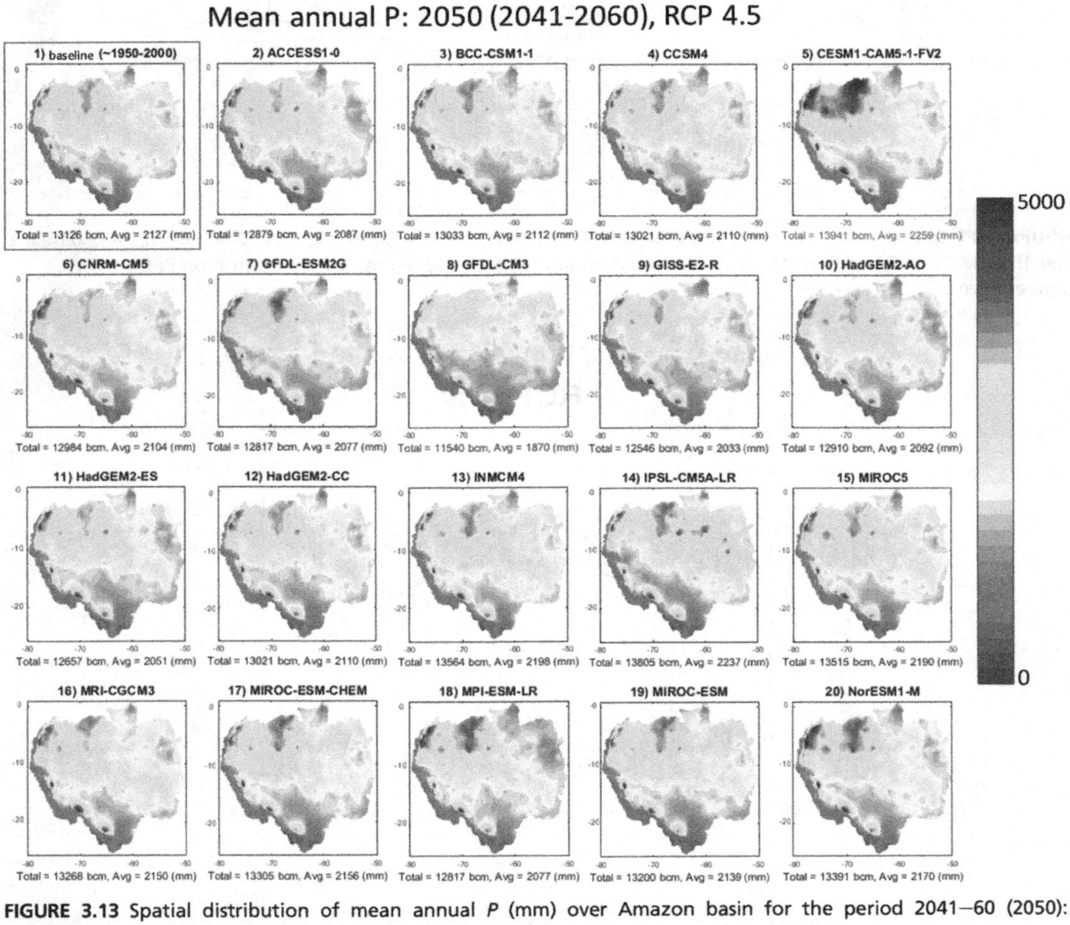

Mean annual P: 2050 (2041-2060), RCP 4.5

FIGURE 3.13 Spatial distribution of mean annual *P* (mm) over Amazon basin for the period 2041–60 (2050): baseline and 19 GCMs, and RCP 4.5. *GCM*, General Circulation Model; *P*, precipitation; *RCP*, Representative Concentration Pathway.

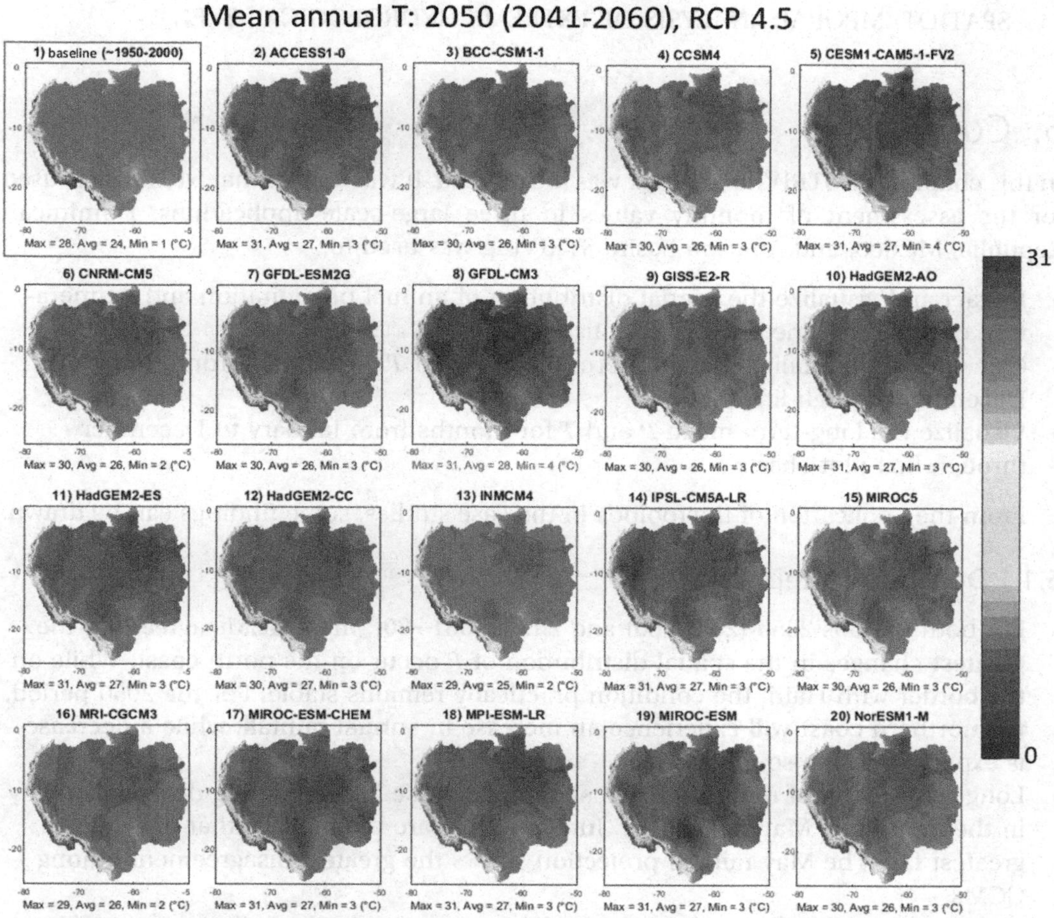

Mean annual T: 2050 (2041-2060), RCP 4.5

1) baseline (~1950-2000) — Max = 28, Avg = 24, Min = 1 (°C)
2) ACCESS1-0 — Max = 31, Avg = 27, Min = 3 (°C)
3) BCC-CSM1-1 — Max = 30, Avg = 26, Min = 3 (°C)
4) CCSM4 — Max = 30, Avg = 26, Min = 3 (°C)
5) CESM1-CAM5-1-FV2 — Max = 31, Avg = 27, Min = 4 (°C)
6) CNRM-CM5 — Max = 30, Avg = 26, Min = 2 (°C)
7) GFDL-ESM2G — Max = 30, Avg = 26, Min = 3 (°C)
8) GFDL-CM3 — Max = 31, Avg = 28, Min = 4 (°C)
9) GISS-E2-R — Max = 30, Avg = 26, Min = 3 (°C)
10) HadGEM2-AO — Max = 31, Avg = 27, Min = 3 (°C)
11) HadGEM2-ES — Max = 31, Avg = 27, Min = 3 (°C)
12) HadGEM2-CC — Max = 30, Avg = 27, Min = 3 (°C)
13) INMCM4 — Max = 29, Avg = 25, Min = 2 (°C)
14) IPSL-CM5A-LR — Max = 31, Avg = 27, Min = 3 (°C)
15) MIROC5 — Max = 30, Avg = 26, Min = 3 (°C)
16) MRI-CGCM3 — Max = 29, Avg = 26, Min = 2 (°C)
17) MIROC-ESM-CHEM — Max = 31, Avg = 27, Min = 3 (°C)
18) MPI-ESM-LR — Max = 31, Avg = 27, Min = 3 (°C)
19) MIROC-ESM — Max = 31, Avg = 27, Min = 3 (°C)
20) NorESM1-M — Max = 30, Avg = 26, Min = 3 (°C)

FIGURE 3.14 Spatial distribution of mean annual *T* (°C) over Amazon basin for the period 2041–60 (2050): baseline and 19 GCMs, and RCP 4.5. *GCM*, General Circulation Model; *RCP*, Representative Concentration Pathway; *T*, temperature.

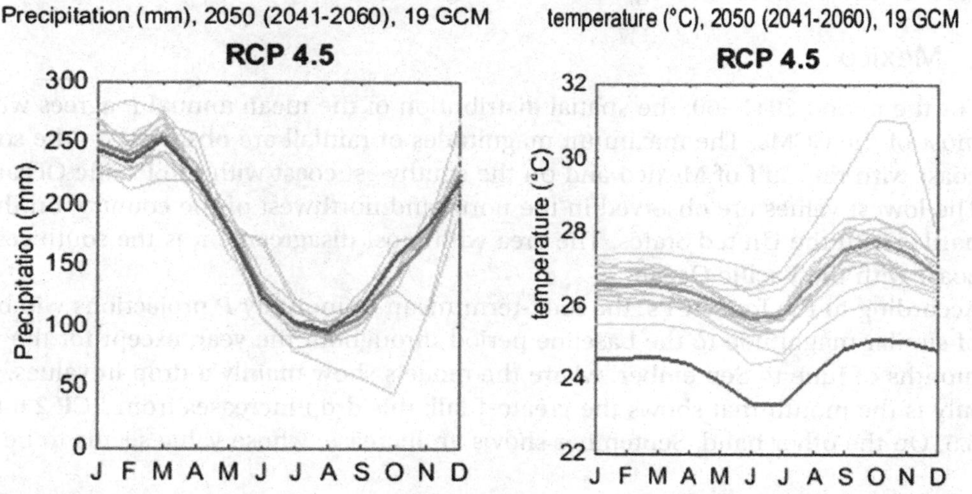

FIGURE 3.15 Amazon basin: long-term mean *P* (*left*) and *T* (*right*) of the baseline period (*solid blue line*) and the 19 RCP 4.5 projections (*solid red line*). Long-term mean *P* of each projection is also displayed for the period 2041–60 (*solid gray lines*). *P*, Precipitation; *RCP*, Representative Concentration Pathway; *T*, temperature.

6. Conclusions

In this chapter the STRIVIng toolbox was introduced. It was shown that STRIVIng is used for the assessment of monthly values in three large-scale applications: Dominican Republic, Mexico, and Amazon basin. STRIVIng was used to:

- Extract and visualize the spatial distribution of annual precipitation and temperature of both baseline and projections.
- Calculate and visualize the long-term mean *P* and *T* for months from January to December through line plots.
- Visualize the long-term mean *P* and *T* for months from January to December through box-plot charts.

From the application of the toolbox in the case studies, some findings can be drawn.

6.1 Dominican Republic

- For both periods 2050 (2041–60) and 2070 (2061–80), most GCMs agree that the greatest changes in the spatial distribution of *P* occur on the north coast. While on the border with Haiti, the condition practically remains stable. For the 2050 period, the northern coast will experience an increase in annual rainfall, while a decrease is expected on the southern coast.
- Long-term mean of monthly *P* will show a decrease in the coming decades, mainly in the months of May to October. June and July are the months that present the greatest fall. The May rainfall projection shows the greatest disagreement among GCMs.
- The country will suffer an increase in the mean annual values of temperature in practically the whole territory for the periods 2041–60 and 2061–80. Only in the central part will the condition remain more stable in both cases.
- An increase of around 2 degrees in the long-term mean of monthly *T* is observed.

6.2 Mexico

- For the period 2041–60, the spatial distribution of the mean annual *P* agrees with most of the GCMs. The maximum magnitudes of rainfall are observed on the south coast with the Gulf of Mexico and on the southwest coast with the Pacific Ocean. The lowest values are observed in the north and northwest of the country, on the border with the United States. The area with most disagreement is the southwest coast with the Pacific Ocean.
- According to the four RCPs, the long-term mean of monthly *P* projections will be of similar magnitude to the baseline period throughout the year, except for the months of June to September, where the models show mainly a drop in values. July is the month that shows the greatest fall; this drop increases from RCP 2.6 to 8.5. On the other hand, September shows an increase, whose value seems to be

lower from route 2.6 to 8.5. July and September are the months with the poorest agreement between GCMs.

- Regarding *T*, results show that most GCMs indicate a rise in temperatures throughout the country for the period 1941–2060. On the coasts and in the north, the greatest values are observed. It is observed that increases in temporal distribution and values of long-term mean *T* projections from January to December are larger than those of the baseline period. The main increases are observed from June to September.

6.3 Amazon Basin

- The spatial distribution of mean annual *P* shows different values along the basin in the baseline period. The largest appear to be in the northwest and center, followed by the rains in the northwest. The lowest are in the southwest part. According to RCP 4.5, most GCMs indicate a change in the spatial distribution of annual rainfall for the period 2041–60, mainly in the northwest of the basin, where the values seem to increase.
- The long-term mean of monthly *P* projections shows that the values from January to July in general remain, but not the values from August to December, where a drop in the magnitude of rainfall is expected. The worst agreement between models is observed in October. This is for the period 2041–60.
- Most GCMs indicate an increase in temperature throughout the territory of the basin. The northern, eastern, and central parts seem to be the most affected. There is also a rise in the southwest part, where the expected average minimum value is 4 degrees. The long-term mean of monthly *T* projections shows a greater disagreement in every month with respect to rainfall projections. In general, projections indicate an increase in *T* in all months. The poorest agreement is observed in October.

The toolbox can be retrieved at https://github.com/hydroinfo4x/STRIVIng.

Acknowledgments

Vitali Díaz thanks the Mexican National Council for Science and Technology (CONACYT) for the study grant 217776/382365. This research was partially financed and supported by the MESCyT and its FONDOCyT program, under the project number 2015-1H1-06 entitled "Impacto del cambio climático en el aprovechamiento sostenible de los recursos hídricos en República Dominicana". We acknowledge WorldClim data developers for processing the climate projections and for making them freely available.

References

Boer, G.J., 2009. Changes in interannual variability and decadal potential predictability under global warming. Journal of Climate 22 (11), 3098–3109. https://doi.org/10.1175/2008JCLI2835.1.

Coumou, D., Rahmstorf, S., 2012. A decade of weather extremes. Nature Climate Change 2 (7), 491–496. https://doi.org/10.1038/nclimate1452.

Fick, S.E., Hijmans, R.J., 2017. WorldClim 2: new 1-km spatial resolution climate surfaces for global land areas. International Journal of Climatology 37 (12), 4302–4315. https://doi.org/10.1002/joc.5086.

Hawkins, E., Sutton, R., 2009. The potential to narrow uncertainty in regional climate predictions. Bulletin of the American Meteorological Society 90 (8), 1095–1107. https://doi.org/10.1175/2009BAMS2607.1.

IPCC, 2014. In: Core Writing Team, Pachauri, R.K., Meyer, L.A. (Eds.), Climate Change 2014: Synthesis Report. Contribution of Working Groups I, II and III to the Fifth Assessment Report of the Intergovernmental Panel on Climate Change. IPCC, Geneva, Switzerland.

Meehl, G.A., Moss, R., Taylor, K.E., Eyring, V., Stouffer, R.J., Bony, S., Stevens, B., 2014. Climate model intercomparisons: preparing for the next phase. Eos 95 (9), 77–78. https://doi.org/10.1002/2014EO090001.

Meinshausen, M., Smith, S.J., Calvin, K., Daniel, J.S., Kainuma, M.L.T., Lamarque, J., et al., 2011. The RCP greenhouse gas concentrations and their extensions from 1765 to 2300. Climatic Change 109 (1), 213–241. https://doi.org/10.1007/s10584-011-0156-z.

Milly, P.C.D., Dunne, K.A., Vecchia, A.V., 2005. Global pattern of trends in streamflow and water availability in a changing climate. Nature 438 (7066), 347–350. https://doi.org/10.1038/nature04312.

Moss, R.H., Edmonds, J.A., Hibbard, K.A., Manning, M.R., Rose, S.K., Van Vuuren, D.P., et al., 2010. The next generation of scenarios for climate change research and assessment. Nature 463 (7282), 747–756. https://doi.org/10.1038/nature08823.

Najafi, M.R., Moradkhani, H., 2015. Multi-model ensemble analysis of runoff extremes for climate change impact assessments. Journal of Hydrology 525, 352–361. https://doi.org/10.1016/j.jhydrol.2015.03.045.

Taylor, K.E., Stouffer, R.J., Meehl, G.A., 2012. An overview of CMIP5 and the experiment design. Bulletin of the American Meteorological Society 93 (4), 485–498. https://doi.org/10.1175/BAMS-D-11-00094.1.

Trenberth, K.E., 2012. Framing the way to relate climate extremes to climate change. Climatic Change 115 (2), 283–290. https://doi.org/10.1007/s10584-012-0441-5.

Wang, L., Ranasinghe, R., Maskey, S., van Gelder, P.H.A.J.M., Vrijling, J.K., 2016. Comparison of empirical statistical methods for downscaling daily climate projections from CMIP5 GCMs: a case study of the Huai River Basin, China. International Journal of Climatology 36 (1), 145–164. https://doi.org/10.1002/joc.4334.

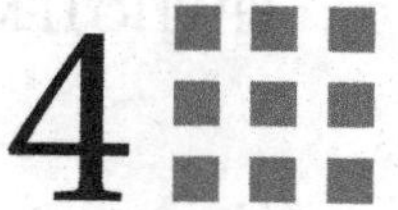

Spatiotemporal Drought Analysis at Country Scale Through the Application of the STAND Toolbox

Vitali Diaz[1,2], Gerald Corzo[1], Henny A.J. Van Lanen[3], Dimitri P. Solomatine[1,2]

[1]*UNESCO-IHE INSTITUTE FOR WATER EDUCATION, DELFT, THE NETHERLANDS;* [2]*WATER RESOURCES SECTION, DELFT UNIVERSITY OF TECHNOLOGY, DELFT, THE NETHERLANDS;* [3]*HYDROLOGY AND QUANTITATIVE WATER MANAGEMENT GROUP, WAGENINGEN UNIVERSITY, WAGENINGEN, THE NETHERLANDS*

1. Introduction

Drought is a natural phenomenon whose impacts generate many economic and human losses (Below et al., 2007; Sheffield and Wood, 2011; Tallaksen and Van Lanen, 2004). Different definitions of drought have been proposed, according to the discipline from where drought is addressed (Bachmair et al., 2016; Mishra and Singh, 2010). The general consensus is that it is an anomaly originating in precipitation and temperature, whose further effects are observed in other components of the hydrological cycle such as soil moisture and runoff, affecting human activities (Tallaksen and Van Lanen, 2004; Van Loon, 2015). However, the spatial extent and duration of drought are not explicitly contemplated in such definitions. Many studies of global models that explore drought require arbitrary steps to define the size (threshold) that can be used to consider an event as extreme or even normal. In this sense, one drought event in a large region can be generalized in a particular time, even if the region does not show drought in its entire area but only in a small part.

To estimate drought magnitude and duration, drought indicators or indices (DIs) are often used. A DI is a mathematical formulation that quantifies the water anomaly in a hydrometeorological variable (Mishra and Singh, 2010, 2011). When the DI is computed over a region in a spatial way, the spatial extent of drought is estimated by performing a spatiotemporal method (e.g., Corzo Perez et al., 2011; Hannaford et al., 2011; Herrera-Estrada et al., 2017; Hisdal and Tallaksen, 2003; Lloyd-Hughes, 2012; Peters et al., 2006; Sheffield et al., 2009; Tallaksen et al., 2009; Tallaksen and Stahl, 2014; Van Huijgevoort et al., 2013; Vicente-Serrano, 2006; Zaidman et al., 2002). Currently, monitoring of drought

Spatiotemporal Analysis of Extreme Hydrological Events. https://doi.org/10.1016/B978-0-12-811689-0.00004-5

magnitude, duration, and area over a region is carried out through drought-monitoring systems. These systems are fed with hydrometeorological variables. The outcomes generated are known as drought-monitoring products. Drought indicator databases are part of this information. These drought-related data perform more detailed analyses on the spatiotemporal development of drought.

It has been highlighted that a better analysis of drought allows the development and implementation of more successful national policies for the mitigation of drought impacts (World Meteorological Organization (WMO), 2006). The WMO also points out that for the reduction of negative drought impacts it is necessary to develop technologies and methods to improve the characterization of drought.

The objective of this chapter is to introduce the Spatio-Temporal ANalysis of Drought (STAND) toolbox. The tool includes two methodologies for the spatiotemporal characterization of drought: (1) Non-Contiguous Drought Area (NCDA) analysis (Corzo Perez et al., 2011), and (2) Drought DuRation, SeveriTy, and Intensity Computing (DDRASTIC). The use of STAND is illustrated with two case studies: Mexico and India. After this introductory section, the NCDA and DDRASTIC methodologies are described, as well as the STAND components. Then, case studies and data are presented. Following this, the results and discussion section is shown.

2. Spatio-Temporal ANalysis of Drought Toolbox

2.1 STAND Analysis

The analysis of drought proposed here can be performed in four main steps aiming at the description of how drought evolves in space (drought area), and how this spatial area changes over time. As mentioned, drought is defined here as an anomaly in the hydrometeorological variable under analysis. This anomaly is detected with the help of a DI. Three types of drought events are defined here based on the way data are processed: time series, spatial, and spatiotemporal events.

Step 1. Temporal analysis: calculation of the region-aggregated drought indicator, as well as computation of the time series events and their duration and deficit.

Step 2. Spatiotemporal analysis: integration of the time series events into aggregated areas of drought, and calculation of the percentage of drought area (PDA) in each time step.

Step 3. Characterization of the spatiotemporal events.

Step 4. Visualization and analysis of results.

2.2 Temporal Analysis (Step 1)

Fig. 4.1 shows a time series of a DI whose values oscillate from −3 to 3. The negative values are associated with the drought anomalies. In grid data, each cell has a DI time series, and over each one the characterization of time series events indicated in this step

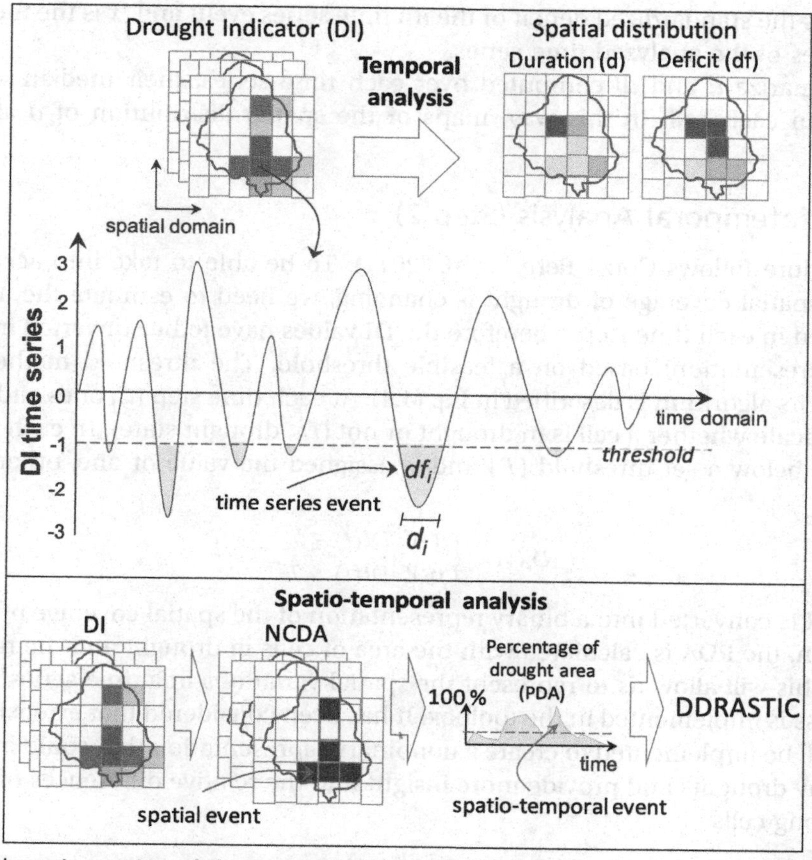

FIGURE 4.1 Schematic overview of the methodologies for drought analysis in the Spatio-Temporal ANalysis of Drought (STAND) toolbox: Non-Contiguous Drought Area (NCDA) analysis and Drought DuRAtion, SeveriTy, and Intensity Computing (DDRASTIC).

is carried out. Following Mckee et al. (1993), a time series event starts at time *ts* when the DI value is below a set threshold (T) and ends at time *te* when DI is above the threshold. Duration (d) and deficit (df) of each ith time series event are computed with Eqs. (4.1) and (4.2), respectively.

$$d_i = te - ts \tag{4.1}$$

$$df_i = \sum_{t=ts}^{te} (DI(t) - T). \tag{4.2}$$

The deficit is standardized and expressed as a percentage with Eq. (4.3):

$$dfs_i = 100 \times df_i/\overline{x} \tag{4.3}$$

where dfs_i is the standardized deficit of the ith time series event and $\bar{x}$ is the mean of the deficit values of the analyzed time series.

To summarize d and df computed over each time series, their median values are calculated in each cell. In this way, maps of the spatial distribution of d and df are obtained.

2.3 Spatiotemporal Analysis (Step 2)

This procedure follows Corzo Perez et al. (2011). To be able to take into account how much the spatial coverage of drought is changing, we need to estimate the amount of area affected in each time step. Therefore the DI values have to be converted into events (binary representation) based on a feasible threshold. The threshold method is well known and its algorithm is described in Eq. (4.4). At each time step (t), ones and zeros are used to indicate whether a cell is in drought or not (D_S, drought state). In each cell, if the DI value is below a set threshold (T) and is assigned the value of one, otherwise zero (Eq. 4.4).

$$D_s(t) = \begin{cases} 1 \text{ if } DI(t) \leq T \\ 0 \text{ if } DI(t) > T \end{cases}. \tag{4.4}$$

After, DI is converted into a binary representation of the spatial coverage of droughts. For a region, the PDA is calculated with the area of cells in drought and the region area (Eq. 4.5). This will allow us to represent the spatial variations in a time series spectrum. This process is implemented in the toolbox. It has been considered that extensions to the method will be implemented to create a nonbinary representation that could integrate in a better way droughts and provide more insight into the relative differences of the index values among cells:

$$PDA(t) = 100 / A_{tot} \cdot \sum_{c=1}^{N} (D_s(t) \cdot A) \tag{4.5}$$

where A is the area of the cell c and A_{tot} is the region area. In this methodology, the magnitude of drought is the PDA value. PDA time series allows the identification of large droughts over a region. Small areas are neglected by applying a second threshold.

2.4 Drought Duration, Severity, and Intensity Computing (Step 3)

This methodology follows Mckee et al. (1993), although hereafter it is extended to be used with the PDA concept used in NCDA. Computing of drought duration (DD) starts at the time step when the PDA lies below the defined threshold. In a PDA series, this threshold is estimated by using the 90th percentile of the low PDA. Therefore the spatiotemporal event starts when the PDA is below the set threshold (T_{PDA}) and ends when it is above the set threshold (Eq. 4.6):

$$DD_j = t_E - t_S \tag{4.6}$$

where j is the jth spatiotemporal event and DD, t_S, and t_E are the drought duration, the start and end time of the event j.

The region under the PDA curve is drought severity (S, expressed as a percentage). This value is a measure of the drought magnitude. S is calculated for each event j with Eq. (4.7):

$$S_j = \sum_{t=t_S}^{t_E} PDA(t). \tag{4.7}$$

The intensity (I) is calculated as the ratio between S and DD (Eq. 4.8):

$$I_j = S_j / DD_j. \tag{4.8}$$

This I_j ratio can be interpreted as the mean value of PDA during the time DD_j. Calculation of DD, S, and I is carried out over the entire time series of the PDA. In this methodology, each triplet DD, S, and I are the characteristics of what here is called a spatiotemporal event. It is important to highlight that if the spatial coverage is not large enough at any given time, there is no drought. This method is an extension of the methodology used in Corzo Perez et al. (2011).

2.5 Visualization and Analysis of Results (Step 4)

The STAND toolbox includes five types of plots designed for the analysis and presentation of outcomes: (1) line plot, (2) scatter plot, (3) area chart, (4) colored array, and (5) image plot. Table 4.1 shows the information that can be displayed in each plot.

2.6 Interpolation of Pointwise Data

For the analysis and modeling of catchments, it is common to have pointwise information of hydrometeorological data. This information has to be interpolated spatially to be used before carrying out the spatiotemporal analysis of drought. The STAND toolbox includes an algorithm based on the inverse distance weighted (IDW), where the missing information is estimated by weighting the neighboring known information. The coefficients of the weighting are inversely proportional to the distances between the point of interest and its closest neighbors.

Table 4.1 Plots for the Visualization of Drought-Related Information in the STAND Toolbox

Plot	Outcome(s)
Line	Drought indicator time series (TA)
Scatter	Duration versus deficit (TA), duration versus severity, duration versus intensity (STA)
Area chart	Time series of percentage of drought area (STA)
Colored array	Time series of percentage of drought area (STA)
Image	Spatial distribution of drought indicator, duration, deficit (TA), and area (STA)

STA, Spatiotemporal analysis; *STAND*, Spatio-Temporal ANalysis of Drought; *TA*, Temporal analysis.

3. Case Study and Data

STAND is applied to two case studies: Mexico and India, two countries prone to drought. For Mexico, Standardized Precipitation Index (SPI) time series calculated by the Servicio Meteorológico Nacional (SMN, National Meteorological Service) were used (Table 4.2). SPI is computed by SMN at 362 stations across the country (Fig. 4.2A). SPI uses only precipitation for its calculation and its methodology is described in Mckee et al. (1993, 1995) and WMO (2012). For India (Fig. 4.2B), grid data from the Standardized Precipitation Evaporation Index (SPEI) Global Drought Monitor were used. The difference between precipitation and evapotranspiration is used to calculate SPEI (Vicente-Serrano et al., 2010). The description of both databases is shown in Table 4.2.

Table 4.2 Summary of Drought Indicator Databases

Database	Repository	Temporal Resolution and Coverage	Spatial Resolution and Coverage	Meteorological Data Source (Input)	Main References of Procedure
SPI	SMN[a]	1–24 months, 1951–2017	362 weather stations, Mexico	SMN weather stations	Mckee et al. (1993) and WMO (2012)
SPEI v2.3	SPEI global[b]	1–48 months, 1901–2013	0.5 degrees, globe	CRU TS 3.23[c] (Harris et al., 2014)	Vicente-Serrano et al. (2010) and Beguería et al. (2014)

[a]SMN-Mexico SPI: Standardized Precipitation Index (SPI) from the Servicio Meteorológico Nacional (SMN, National Meteorological Service)-Mexico (http://smn.cna.gob.mx/es/climatologia/monitor-de-sequia/spi).
[b]SPEI global: Standardized Precipitation Evaporation Index (SPEI) Global Drought Monitor (https://digital.csic.es/handle/10261/128892).
[c]Climatic Research Unit time series 3.23.

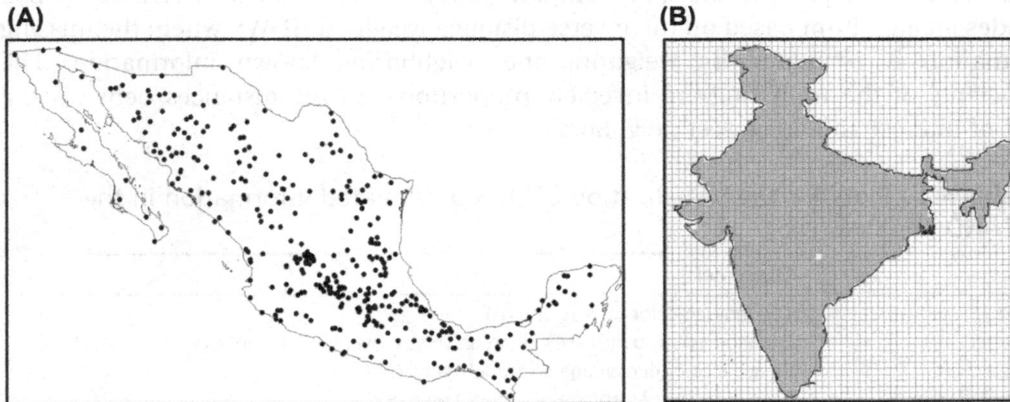

FIGURE 4.2 (A) 362 weather stations where Standardized Precipitation Index (SPI) is calculated by Servicio Meteorológico Nacional (SMN). (B) Region where 0.5-degree Standardized Precipitation Evaporation Index (SPEI) data were used to analyze drought over India.

4. Experiment Setup

For Mexico and India, the experiments were carried out according to the STAND methodology (Section 2). In both cases, for SPI and SPEI the aggregation period of 6 months was chosen to illustrate the application of STAND. This is indicated with SPI6 and SPEI6, respectively. For the case of Mexico, data were first spatially interpolated to obtain the DI values in a grid with a 0.01 degree resolution. This resolution was considered appropriate to reduce the overlap of points in the same cell.

Temporal analysis was performed to identify regions of critical variations of temporal events through their duration and deficit. Also, the time series of country-average SPI6 and SPEI6 were calculated. Afterward, through NCDA analysis, the drought areas and their percentages (PDAs) were calculated. In both cases, for each cell and each time step, the DI value of −1 was set as the threshold to consider an anomaly as drought or not through ones and zeros, respectively. Following DDRASTIC methodology, durations, severities, and intensities of the spatiotemporal events were calculated (Section 2.4). In the PDA series, the 90th percentile of the low PDA was used as threshold.

Results of the STAND methodology were analyzed through the graphic elements that are included in the toolbox. Results were also compared with local information reported in the Emergency Events Database (EM-DAT, Guha-Sapir, 2018). The most severe drought events were also analyzed.

5. Results and Discussion

5.1 Mexico

5.1.1 Step 1: Temporal Analysis

Fig. 4.3 shows the range of SPI6 values of all the cells for the period 1951–2017 (67 years). Values from the SPI6 database fluctuate between −6 and 6. Country-average SPI6 is displayed with a colored line. We can identify that in some decades there are country-average SPI6 values lower than threshold −1, for example, the 1970s, 1980s, 1990s, 2000s, and the current decade (2010s). These low peaks indicate months where on average the country presented SPI values lower than −1, which point out large portions of the territory in drought. These results are consistent with that reported in the

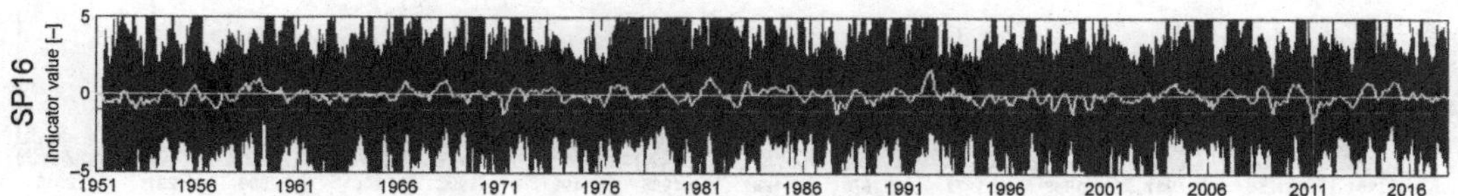

FIGURE 4.3 SPI6 time series per cell. *Colored line* indicates the country-average SPI6 of Mexico.

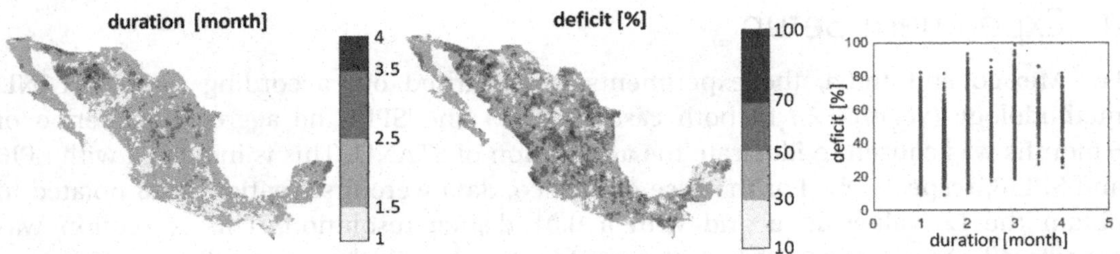

FIGURE 4.4 Median duration and deficit of time series events in Mexico based on SPI6 (1951–2017).

EM-DAT, where, for example, the drought of 2011 is indicated as one of the most impactful in Mexico (Guha-Sapir, 2018).

The spatial distribution of median duration (left) and deficit (center) of the time series events are presented in Fig. 4.4. It is observed that northern Mexico is the zone with the highest density of events with the longest duration and deficit. This was expected since in northern Mexico the arid climate predominates and the monthly precipitation values are the lowest in the country. The maximum of median durations in the region appears to be 4 months, and shorter events are present in central and southern regions. The duration of 1 month seems to be along the whole country showing that these short durations could be noise and should be removed. No linear relationship between median duration and deficit is observed in Fig. 4.4 (right).

5.1.2 Step 2: Spatiotemporal Analysis

Fig. 4.5 presents PDA series calculated for Mexico. It is observed that throughout the decades the spatial extent of droughts fluctuates from zero, reaching in some years peaks even greater than 50%. The PDAs sometimes grow over the months and then decrease, as in 1952–53. On other occasions the PDAs grow suddenly as in 1971 and 2011. In others, the maximum PDAs are between low values as in 1998–2000. The maximum PDA was presented in 2011, possibly indicating that the spatial extent of the drought is an important factor in its impact, since the drought reported in 2011 was one of the worst in Mexico.

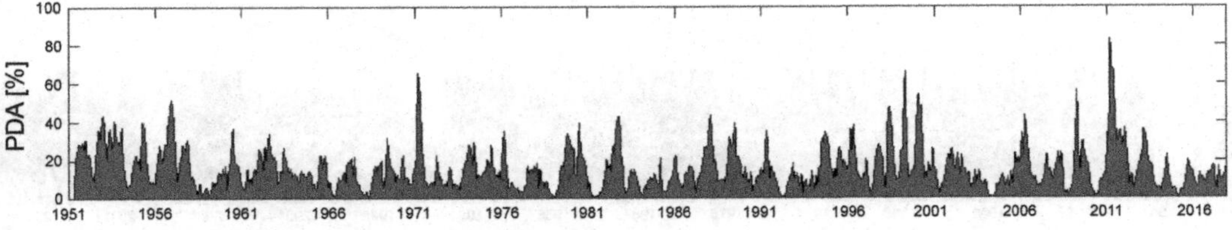

FIGURE 4.5 Area chart with percentages of drought area in Mexico based on SPI6 (1951–2017).

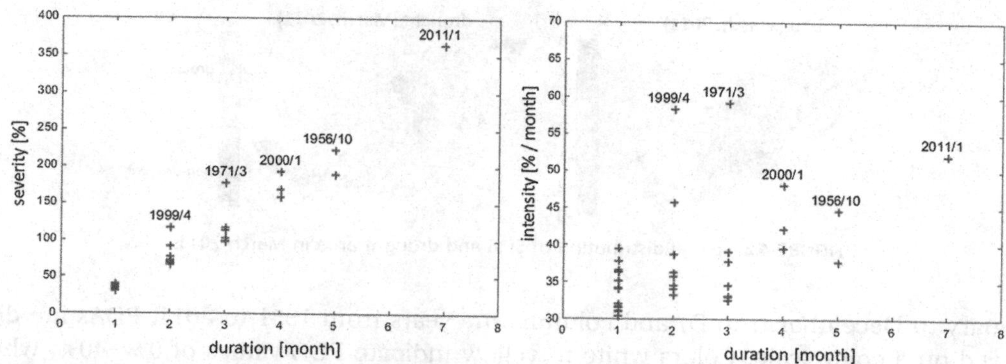

FIGURE 4.6 Duration versus severity (*left*) and duration versus intensity (*right*) of spatiotemporal events in Mexico based on DDRASTIC methodology. Some of the most severe events are indicated.

5.1.3 Step 3: Characterization of Spatiotemporal Events

Duration and severity of spatiotemporal events calculated from PDA series are shown in Fig. 4.6 (left). The year and month of the start of each event are indicated as well in the figure. An almost linear relationship between duration and severity is observed. The event with the longest duration is that of 2011, which started in January and ended in July, lasting 7 months. Four other events with important values of severity are shown: 1956, 1971, 1999, and 2000. Duration versus intensity is presented in Fig. 4.6 (right). It is observed that the event of 1971 presents the maximum intensity. With a duration of 3 months, this event reached an intensity of around 60%/month, i.e., for 3 months the average PDA was 60%. In the event of 2011, average PDA was a little over 50% for 7 months.

5.1.4 Step 4: Visualization

In previous paragraphs, results of the STAND methodology have been presented using graphic elements that are included in the toolbox. In this section, other plots that can be used to display and analyze spatial and spatiotemporal events are shown. PDAs are displayed now in a colored array in Fig. 4.7. Rows make reference to the months of

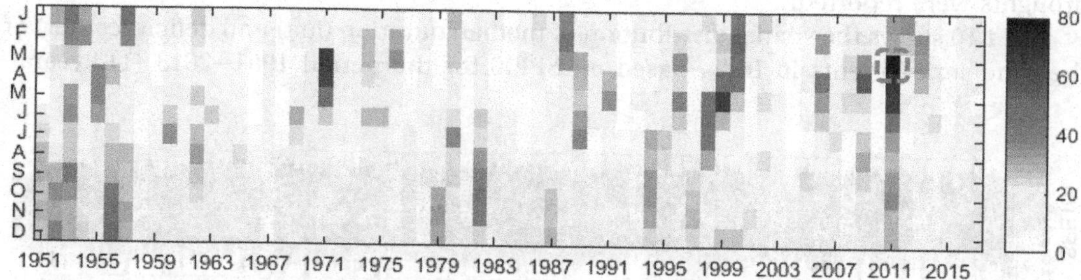

FIGURE 4.7 Percentage of drought area (PDA) for the period 1951–2017 based on SPI6. It is indicated that the highest PDA is in March 2011 (details in Fig. 4.8).

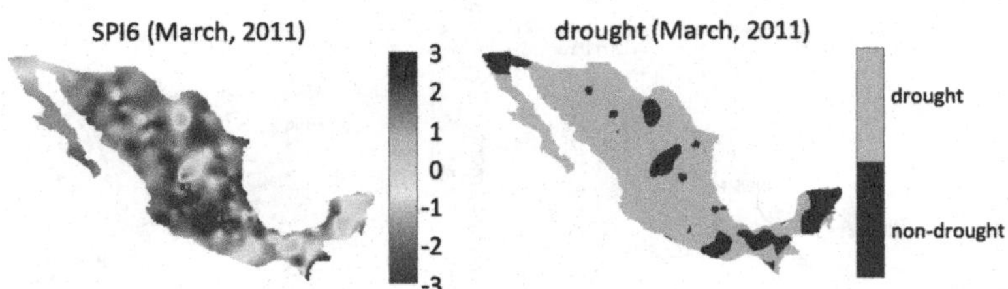

FIGURE 4.8 Spatial distribution of SPI6 and drought area in March 2011.

January to December (J to D) and columns the years from 1951 to 2017. PDAs are displayed on a color scale. Colors white to yellow indicate PDA values of 0%–40%, while orange to dark red indicate PDA values from 40% to 80%. Whitish sections point out periods with small or no PDA values. It is observed that, in general, PDA values greater than 50% occur in the first semester (January to June). The years of 1956, 1957, 1971, 1998, 1999, 2000, 2009, and 2011 present monthly PDAs greater than 50%. The maximum PDA corresponds to March 2011. In this year, values greater than 50% begin in March, continue, and end in June. Fig. 4.8 shows the spatial distribution of SPI6 and the drought area calculated for March 2011. The effect of the interpolation by the IDW method on the spatial distribution of the SPI values is observed. The shape and extent of the drought areas seem to depend on the interpolation method used. Thus it is advisable to perform a sensitivity analysis to study the effect of the chosen interpolation method on the calculation of the PDA.

5.2 India

5.2.1 Step 1: Temporal Analysis

The range of SPEI6 in each cell over India is presented in Fig. 4.9. Also the country-average SPEI6 is displayed (colored line). Values of SPEI6 databases oscillate between −3 and 3. In some decades, country-average SPEI6 values lower than −1 are observed, e.g., the 1900s, 1910s, 2000s, and the current decade. These low peaks point out than in those decades the country had large areas in drought. This is consistent with Bhalme and Mooley (1980) and Guha-Sapir, (2018), where in some of these years, severe droughts were reported.

Fig. 4.10 shows the spatial distribution of median duration (left) and deficit (center) of the time series events in India based on SPEI6 for the period 1901–2013 (113 years).

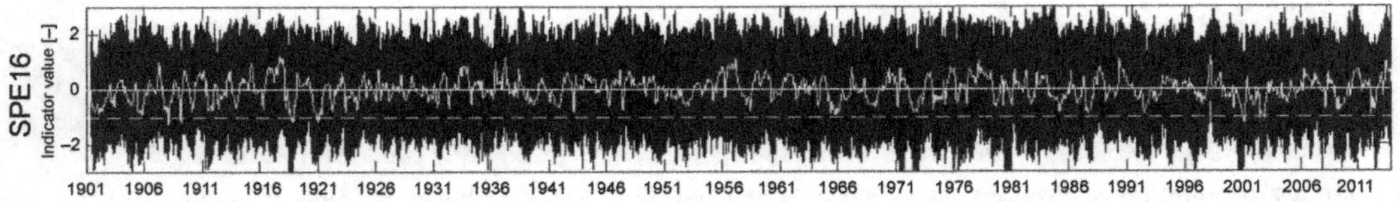

FIGURE 4.9 SPEI6 time series per cell. *Colored line* indicates the country-average SPEI6 of India.

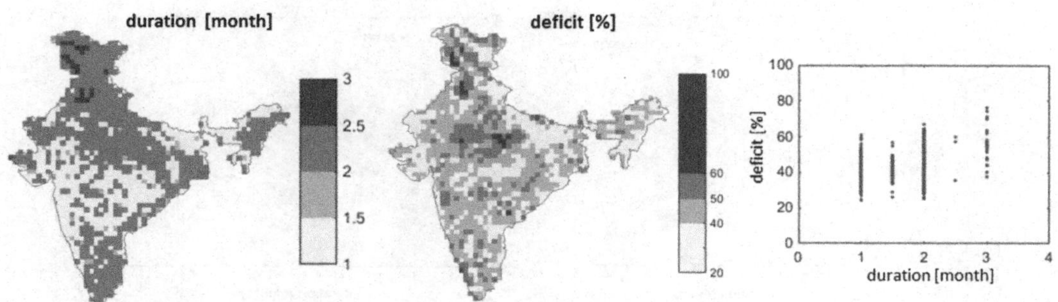

FIGURE 4.10 Median duration and deficit of time series events in India based on SPEI6 (1901–2013).

The longest events are mainly in the north, northeast, and south. On the other hand, deficit is greater in central and north India. The maximum of median durations is observed to be 3 months, and most events have a duration between 2 and 2.5 months. The relationship between median duration and the deficit shown in Fig. 4.10 (right) is not linear, but it is observed that there is a slight positive tendency in the deficit increase as the duration grows.

5.2.2 Step 2: Spatiotemporal Analysis

The time series of PDA calculated for India is shown in Fig. 4.11. The presence of peaks greater than 40% is observed throughout the series. The event of 2001 is the maximum reaching a value of 72%. Other important events are those of 1920, 1972, and 2002. In some of these years, the most severe droughts occurred in India according to Bhalme and Mooley (1980) and Guha-Sapir (2018).

5.2.3 Step 3: Characterization of Spatiotemporal Events

Fig. 4.12 (left) presents duration versus severity of spatiotemporal events calculated from PDAs. The year and month of the start of each event are also indicated in the figure. The relationship between duration and severity seems to be linear. In 1972 the longest event was presented with 9 months of duration, starting in April and ending in December. On the other hand, the event with the maximum severity is that of 1920, which began in October and ended in May 1921, lasting 8 months. Duration versus intensity is presented

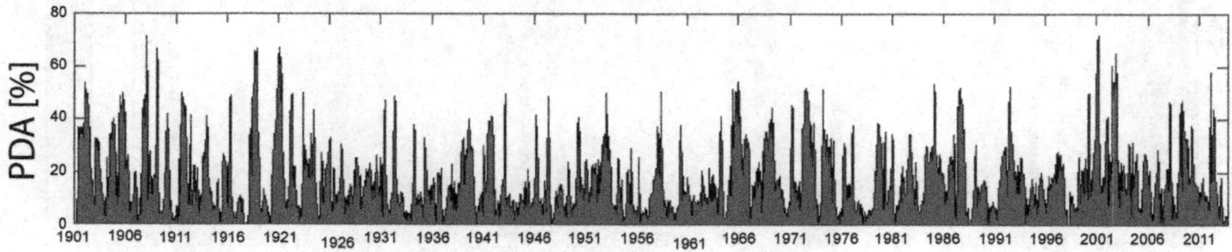

FIGURE 4.11 Area chart with percentages of drought area in India based on SPEI6 (1901–2013).

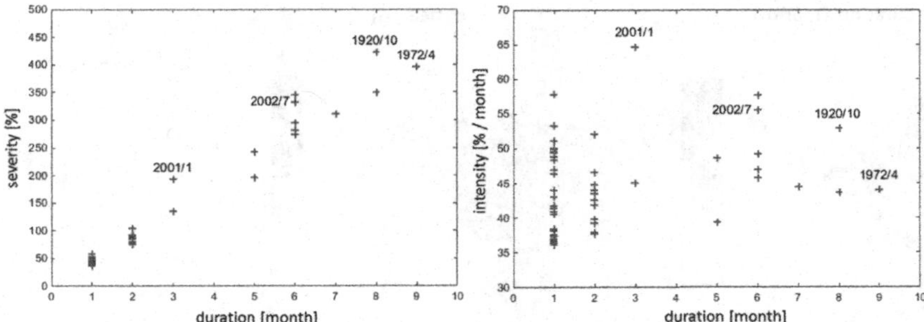

FIGURE 4.12 Duration versus severity (*left*) and duration versus intensity (*right*) of spatiotemporal events in India based on DDRASTIC methodology. Some of the most severe events are indicated.

in Fig. 4.12 (right). It is observed that the event of 2001 is the one of greater intensity with a value of around 65%/month. An important event is that of 1920, which for 8 months averages a value of just over 50% of PDA.

5.2.4 Step 4: Visualization

Outcomes of the STAND methodology were displayed in the previous sections. In this section, other plots to observe and analyze spatial and spatiotemporal events are also shown. In Fig. 4.13 the PDAs are displayed for the period 1901–2013 (113 years). The color scale indicates the percentages. The whitish and yellow spaces indicate periods with null and low values of drought, respectively. Colors orange to dark red refer to values of 50%–100%. Different patterns are observed in the PDA array of Fig. 4.13. For example, a series of successive PDAs greater than 50% that begin in July and last until December are identified, such as the years 1918, 1972, and 2002, to name a few. In some years, the succession of PDAs continues until the following year, for example, 2001–02.

Fig. 4.14 shows how drought areas change in space and time for the years 1920, 1921, 1972, 1973, and 2000–03. This is done to analyze the events of 1920, 1972, 2001, and 2002 presented in the previous section. In the case of India, the PDA threshold is around 36% for the DDRASTIC methodology. Based on this methodology, the 1920 event began in

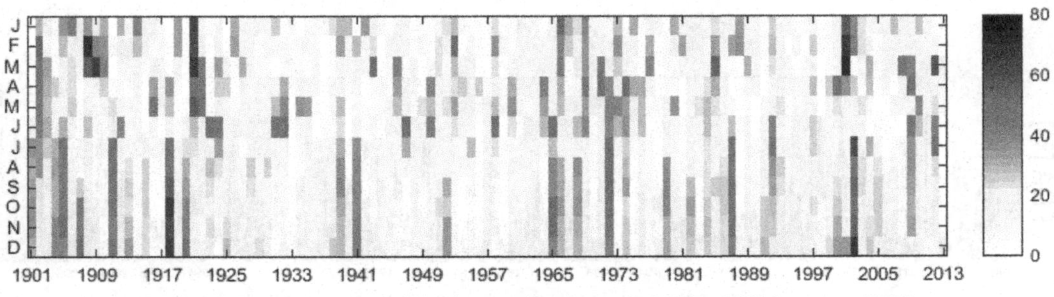

FIGURE 4.13 Percentage of drought area for the period 1901–2013 based on SPEI6.

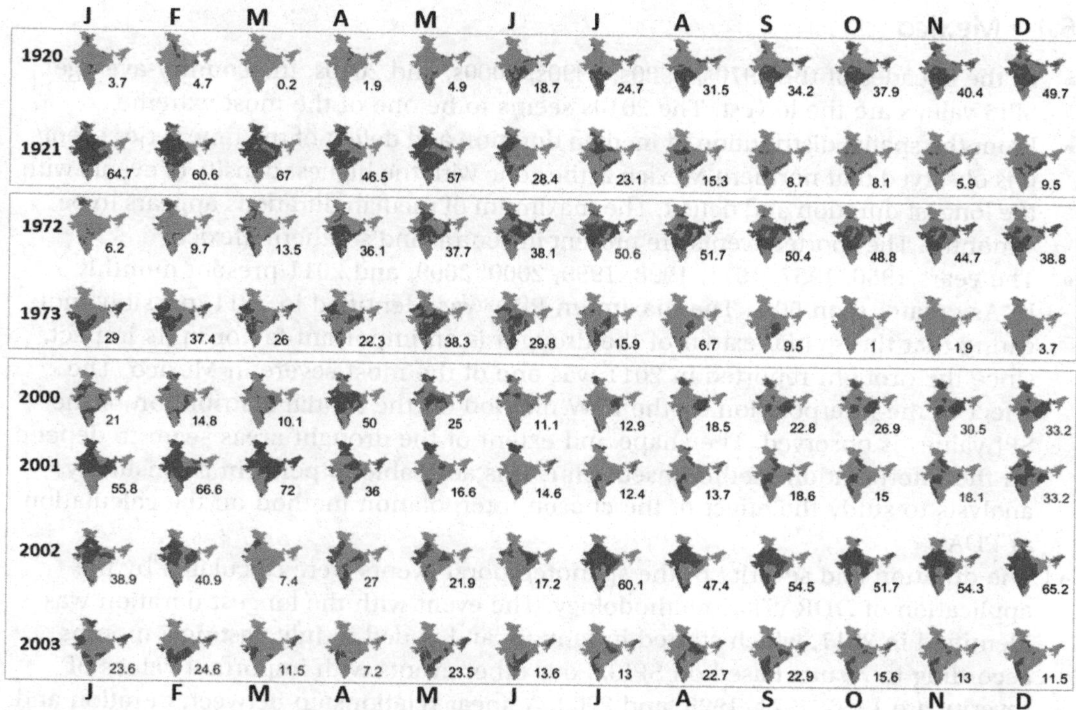

FIGURE 4.14 Spatiotemporal variability of drought areas in India of selected years based on SPEI6. The percentage of drought area is indicated.

October and ended in May 1921, lasting 8 months. For this period, two drought areas are observed mainly in Fig. 4.14, which merged in January 1921. With respect to 1972, the event started in April and ended in December of the same year (9 months). In this case, one area is observed mainly in the central part. The events of 2001 and 2000 seem to be composed of a single drought area as well. The year 2001 began in January and ended in March (3 months), while 2002 began in July and ended in December (6 months). The maximum value of PDA was presented in March 2001 with 72%.

6. Conclusions

The STAND toolbox was introduced and its use was illustrated through two case studies. Two methodologies for the characterization of drought were included within STAND: (1) NCDA analysis and (2) DDRASTIC. It is concluded that the STAND toolbox assisted in the realization of spatiotemporal analysis of drought due to its modules to prepare, process, and visualize drought-related information.

In the case studies for Mexico and India, some findings can be pointed out based on the application of STAND methodology.

6.1 Mexico

- In the decades of the 1970s, 1980s, 1990s, 2000s, and 2010s, the country-average SPI6 values are the lowest. The 2010s seems to be one of the most extreme.
- From the spatial distribution of median duration and deficit of the time series events, it is observed that northern Mexico is the zone with the highest density of events with the longest duration and deficit. The maximum of median durations appears to be 4 months. The shorter events are present in central and southern Mexico.
- The years 1956, 1957, 1971, 1998, 1999, 2000, 2009, and 2011 present monthly PDAs greater than 50%. The maximum PDA was identified in 2011, possibly indicating that the spatial extent of the drought is an important factor in its impact, since the drought reported in 2011 was one of the most severe in Mexico. The effect of the interpolation by the IDW method on the spatial distribution of the SPI values is observed. The shape and extent of the drought areas seem to depend on the interpolation method used. Thus it is advisable to perform a sensitivity analysis to study the effect of the chosen interpolation method on the calculation of PDA.
- The duration and severity of the spatiotemporal events were calculated by the application of DDRASTIC methodology. The event with the longest duration was identified in 2011, which started in January and ended in July, lasting 7 months, according to results based on SPI6. Four other events with important values of severity are 1956, 1971, 1999, and 2000. A linear relationship between duration and severity is observed. On the other hand, the event of 1971 presented the maximum intensity. With a duration of 3 months, this event reached an intensity of around 60%/month, i.e., for 3 months the average PDA was 60%. In the event of 2011, average PDA was a little over 50% for 7 months.

6.2 India

- Decades with the lowest country-average SPEI6 values were the 1900s, 1910s, 2000s, and 2010s. This agrees with what is reported.
- The spatial distribution of median duration and deficit of the time series events show that the longest events are mainly in the north, northeast, and south. On the other hand, deficit is greater in central and north India. The maximum of median durations is observed to be 3 months, and most events have a duration of between 2 and 2.5 months.
- The maximum PDA was found in 2001. It reached a value of 72%. Other important events are 1920, 1972, and 2002. In some of these years, the most severe droughts occurred in India according to the report.
- From the application of DDRASTIC methodology, it is observed that duration and severity of the spatiotemporal events have a linear relationship. In 1972 the longest event was presented with 9 months of duration, starting in April and ending in December. The event with the maximum severity was that of 1920, which began in

October and ended in May 1921, lasting 8 months. On the other hand, the event of 2001 was the one of greater intensity with a value of around 65%/month.

- Analysis of the spatiotemporal events of 1920, 1972, 2001, and 2002, where the most severe are presented according to DDRASTIC methodology, indicates that the event began in October 1920 and ended in May 1921, lasting 8 months. For this period, two drought areas were mainly observed, which merged in January 1921. With respect to 1972, the event started in April and ended in December of the same year (9 months). In this case, one area is observed, mainly in the central part. The events of 2001 and 2000 seem to be composed of a single drought area as well. The year 2001 began in January and ended in March (3 months), while the one in 2002 began in July and ended in December (6 months). The maximum value of PDA was presented in March 2001 with 72%.

The STAND toolbox is available at www.researchgate.net/project/STAND-Spatio-Temporal-ANalysis-of-Drought. A further step in this research is the design and development of charts to show and analyze the variability of the spatiotemporal characteristics of drought. These plots are based on radial diagrams whose methodology and code can be obtained in the same repository shown earlier.

Acknowledgments

Vitali Díaz thanks the Mexican National Council for Science and Technology (CONACYT) for study grant 217776/382365. HvL is supported by the H2020 ANYWHERE project (Grant Agreement No. 700099). We thank the SMN for providing SPI data for Mexico.

References

Bachmair, S., Stahl, K., Collins, K., Hannaford, J., Acreman, M., Svoboda, M., et al., 2016. Drought indicators revisited: the need for a wider consideration of environment and society. Wiley Interdisciplinary Reviews: Water 3 (4), 516–536. https://doi.org/10.1002/wat2.1154.

Beguería, S., Vicente-Serrano, S.M., Reig, F., Latorre, B., 2014. Standardized precipitation evapotranspiration index (SPEI) revisited: parameter fitting, evapotranspiration models, tools, datasets and drought monitoring. International Journal of Climatology 34 (10), 3001–3023. https://doi.org/10.1002/joc.3887.

Below, R., Grover-Kopec, E., Dilley, M., 2007. Documenting drought-related disasters: a global reassessment. The Journal of Environment & Development 16 (3), 328–344. https://doi.org/10.1177/1070496507306222.

Bhalme, H.N., Mooley, D.A., 1980. Large-scale droughts/floods and monsoon circulation. Monthly Weather Review 108 (8), 1197–1211. https://doi.org/10.1175/1520-0493(1980)108<1197:LSDAMC>2.0.CO;2.

Corzo Perez, G.A., Van Huijgevoort, M.H.J., Voß, F., Van Lanen, H.A.J., 2011. On the spatio-temporal analysis of hydrological droughts from global hydrological models. Hydrology and Earth System Sciences 15 (9), 2963–2978. https://doi.org/10.5194/hess-15-2963-2011.

Guha-Sapir, D., January 14, 2018. EM-DAT: The Emergency Events Database. Université catholique de Louvain (UCL) – CRED, Brussels, Belgium. Retrieved from: www.emdat.be.

Hannaford, J., Lloyd-Hughes, B., Keef, C., Parry, S., Prudhomme, C., 2011. Examining the large-scale spatial coherence of European drought using regional indicators of precipitation and streamflow deficit. Hydrological Processes 25 (7), 1146–1162. https://doi.org/10.1002/hyp.7725.

Harris, I., Jones, P.D., Osborn, T.J., Lister, D.H., 2014. Updated high-resolution grids of monthly climatic observations – the CRU TS3.10 dataset. International Journal of Climatology 34 (3), 623–642. https://doi.org/10.1002/joc.3711.

Herrera-Estrada, J.E., Satoh, Y., Sheffield, J., 2017. Spatio-temporal dynamics of global drought. Geophysical Research Letters 2254–2263. https://doi.org/10.1002/2016GL071768.

Hisdal, H., Tallaksen, L.M., 2003. Estimation of regional meteorological and hydrological drought characteristics: a case study for Denmark. Journal of Hydrology 281 (3), 230–247. https://doi.org/10.1016/S0022-1694(03)00233-6.

Lloyd-Hughes, B., 2012. A spatio-temporal structure-based approach to drought characterisation. International Journal of Climatology 32 (3), 406–418. https://doi.org/10.1002/joc.2280.

Mckee, T.B., Doesken, N.J., Kleist, J., January 1993. The relationship of drought frequency and duration to time scales. In: AMS 8th Conference on Applied Climatology, pp. 179–184. Article id: 10490403.

McKee, T.B., Doesken, N.J., Kleist, J., 1995. Drought monitoring with multiple time scales. In: Conference on Applied Climatology.

Mishra, A.K., Singh, V.P., 2010. A review of drought concepts. Journal of Hydrology 391 (1–2), 202–216. https://doi.org/10.1016/j.jhydrol.2010.07.012.

Mishra, A.K., Singh, V.P., 2011. Drought modeling – a review. Journal of Hydrology 403 (1–2), 157–175. https://doi.org/10.1016/j.jhydrol.2011.03.049.

Peters, E., Bier, G., van Lanen, H.A.J., Torfs, P.J.J.F., 2006. Propagation and spatial distribution of drought in a groundwater catchment. Journal of Hydrology 321 (1–4), 257–275. https://doi.org/10.1016/j.jhydrol.2005.08.004.

Sheffield, J., Wood, E.F., 2011. In: Earthscan, P. (Ed.), Drought: Past Problems and Future Scenarios (London).

Sheffield, J., Andreadis, K.M., Wood, E.F., Lettenmaier, D.P., 2009. Global and continental drought in the second half of the twentieth century: severity-area-duration analysis and temporal variability of large-scale events. Journal of Climate 22 (8), 1962–1981. https://doi.org/10.1175/2008JCLI2722.1.

Tallaksen, L.M., Stahl, K., 2014. Spatial and temporal patterns of large-scale droughts in Europe: model dispersion and performance. Geophysical Research Letters 41 (2), 429–434. https://doi.org/10.1002/2013GL058573.

Tallaksen, L.M., Van Lanen, H.A.J., 2004. In: Tallaksen, L.M., Van Lanen, H.A.J. (Eds.), Hydrological Drought – Processes and Estimation Methods for Streamflow and Groundwater, Developments in Water Sciences 48. Elsevier B.V, The Netherlands.

Tallaksen, L.M., Hisdal, H., Van Lanen, H.A.J., 2009. Space-time modelling of catchment scale drought characteristics. Journal of Hydrology 375 (3–4), 363–372. https://doi.org/10.1016/j.jhydrol.2009.06.032.

Van Huijgevoort, M.H.J., Hazenberg, P., van Lanen, H.a.J., Teuling, a.J., Clark, D.B., Folwell, S., Gosling, S.N., Hanasaki, N., Heinke, J., Koirala, S., Stacke, T., Voss, F., Sheffield, J., Uijlenhoet, R., 2013. Global Multimodel Analysis of Drought in Runoff for the Second Half of the Twentieth Century. Journal of Hydrometeorology 14 (5), 1535–1552. https://doi.org/10.1175/JHM-D-12-0186.1.

Van Loon, A.F., 2015. Hydrological drought explained. Wiley Interdisciplinary Reviews: Water 2 (4), 359–392. https://doi.org/10.1002/wat2.1085.

Vicente-Serrano, S.M., Beguería, S., López-Moreno, J.I., 2010. A multiscalar drought index sensitive to global warming: the standardized precipitation evapotranspiration index. Journal of Climate 23 (7), 1696–1718. https://doi.org/10.1175/2009JCLI2909.1.

Vicente-Serrano, S.M., 2006. Differences in spatial patterns of drought on different time scales: an analysis of the Iberian Peninsula. Water Resources Management 20 (1), 37–60. https://doi.org/10.1007/s11269-006-2974-8.

World Meteorological Organisation, 2006. Drought Monitoring and Early Warning: Concepts, Progress and Future Challenges. WMO – No. 1006. Retrieved from: http://www.wamis.org/agm/pubs/brochures/WMO1006e.pdf.

World Meteorological Organisation, 2012. Standardized Precipitation Index User Guide. WMO – No. 1090. Retrieved from: http://www.wamis.org/agm/pubs/SPI/WMO_1090_EN.pdf.

Zaidman, M.D., Rees, H.G., Young, A.R., 2002. Spatio-temporal development of streamflow droughts in north-west Europe. Hydrology and Earth System Sciences 6, 733–751. https://doi.org/10.5194/hess-6-733-2002.

Spatiotemporal Analysis of Extreme Rainfall Events Using an Object-Based Approach

Miguel Laverde-Barajas[1,2], Gerald Corzo[1], Biswa Bhattacharya[1], Remko Uijlenhoet[3], Dimitri P. Solomatine[1,2]

[1]IHE DELFT INSTITUTE OF WATER EDUCATION, DELFT, THE NETHERLANDS; [2]DELFT UNIVERSITY OF TECHNOLOGY, WATER RESOURCES SECTION, DELFT, THE NETHERLANDS; [3]HYDROLOGY AND QUANTITATIVE WATER MANAGEMENT GROUP, WAGENINGEN UNIVERSITY, WAGENINGEN, THE NETHERLANDS

1. Introduction

Heavy rainfall is considered to be one of the main causes of flood hazard (Vörösmarty et al., 2013; Wilhelmi and Morss, 2013). During the last decade, events caused by heavy rainfall have been responsible for more than 1 billion people being affected and causing 80% of total global economic damage of all natural hazards (Knight, 2011). Characteristics of rainfall events such as intensity, duration, location, and spatial extension play a fundamental role in determining the level of damage associated with floods. In the era of big data, advances in earth observation systems have provided a detailed description of rainfall fields for many hydrological applications. Despite the availability of high-resolution rainfall data, numerous researchers have shown that characterizing small-scale variability of rainfall patterns is still challenging (Grayson and Blöschl, 2001). Grid-based studies are traditionally characterized by the use of continuous verification statistics to analyze the rainfall dynamic and evaluate product quality. While these measurements provide useful information in terms of correlation, metrics do not consider important intrinsic features of rainfall data such as location, volume, and type of event. Nowadays, high-resolution satellite information has become an important input in water management, flood monitoring, and forecast systems. It is important to understand the capabilities and limitations for estimating the rainfall structure in space and time.

Object-based methods are an alternative approach to analyze rainfall. This methodology evaluates rainfall estimation based on the structural properties of rainfall fields. Based on this method, a rainfall event can be described through the summary of all

Spatiotemporal Analysis of Extreme Hydrological Events. https://doi.org/10.1016/B978-0-12-811689-0.00005-7

intrinsic attributes and statistics representing space and/or time. Different authors have used object-based methods to evaluate the performance of satellite-based products (Davis et al., 2009a; Ebert and McBride, 2000; Li et al., 2015; Skok et al., 2009). They demonstrated how features of rainfall characteristics extracted by object-based methods such as shape orientation and size allow a better description of rainfall.

One important challenge in object-based methods arising from the description of the rainfall fields is analysis using a multidimensional approach. This research describes the innovative application of a spatiotemporal object-based method to analyze extreme rainfall events at the catchment scale, using a subtropical catchment of the Tiete River, Brazil, as an example for identification of extreme rainfall events and the evaluation of near-real time (NRT) satellite-based products.

2. Object-Based Analysis

In nature, an object can be identified through the summary of all intrinsic characteristics. A plant, for instance, could be identified in an image by observing characteristics such as color, size, and shape, etc. Similar to our eyes registering all this information to decide what type of plant we are observing, object-based methods analyze the information of a physical process to understand its characteristics and examine their relationships. Object-based or object-oriented analysis is defined as the study of the statistics of the population of objects. Analysis could be incorporated into a group of images or gridded data to describe an object with low computational complexity (Blaschke et al., 2004).

In the literature, object-based methods have been used for many purposes, from investigating terrain morphology and detecting its changes (Blaschke, 2010; Desclée et al., 2006; Yu et al., 2010) to evaluating the performance of climatic models (Davis et al., 2009a; Grams et al., 2006). In atmospheric sciences, object-based models are widely used for monitoring and nowcasting precipitation systems (Dixon and Wiener, 1993; Han et al., 2009; Johnson et al., 1998) and the spatial verification of weather products (Ahijevych et al., 2009; Brown et al., 2004; Davis et al., 2009a; Ebert and McBride, 2000). In the latter case, multiple spatial verification methods have been proposed to evaluate model accuracy in terms of spatial pattern, intensity, and displacement, such as the contiguous rain area (CRA) (Fig. 5.1) (Ebert and McBride, 2000), object-based diagnostic evaluation (MODE) (Davis et al., 2009a), and structure, amplitude, and location (SAL) (Wernli et al., 2008). Gilleland et al. (2009) discussed the capabilities of spatial object-based verification methods in proving information on structure errors of rainfall estimation at multiple scales.

Advances in high-resolution satellite rainfall detection have led to employing these methods for evaluating the accuracy of satellite rainfall products at large scales. For instance, Skok et al. (2009) implemented the MODE method to analyze the properties and spatial distribution of the rainfall systems from the Precipitation Estimation from

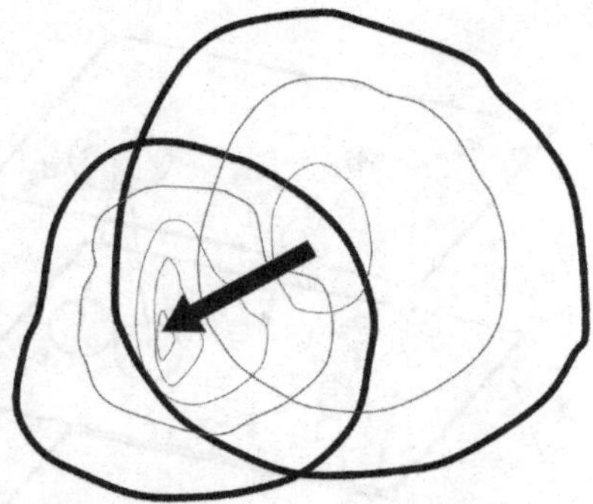

FIGURE 5.1 Schematic diagram of a contiguous rain area. The arrow shows the level of displacement between the observed and predicted rainfall fields. *From Ebert, E.E., 2007. Methods for verifying satellite precipitation estimates. In: Measuring Precipitation from Space, Springer, 345–356.*

Remotely Sensed Information using Artificial Neural Networks (PERSIANN, Sorooshian et al. (2000)) and the Tropical Rainfall Measuring Mission (TRMM 3B42, Huffman (2007)) over the equatorial Pacific Ocean. Additionally, Demaria et al. (2011) used the CRA method to identify systematic errors in South America using TRMM, PERSIANN, and the climate prediction center (CPC) morphing technique (CMORPH, Joyce et al. (2004)). Li et al. (2015) developed an object-based approach to validate satellite-based rainfall products against ground observations. This method was later applied by Li et al. (2016) for evaluating three high-resolution satellite precipitation products (PERSIANN, CMORPH, and TRMM) in the United States. One of the main advantages of these methods in comparison to traditional gridded-based approaches was the incorporation of new information about errors in shape, orientation, and displacement into rainfall analysis.

3. Spatiotemporal Object-Based Methods

Object-based methods have in common a 2D approach to identify rainfall objects in space (latitude, longitude). The evolution of rainfall systems in time is analyzed simply through a connectivity or "tracking" function (e.g., overlapping, centroid displacement). Once the attributes of two pairs of consecutive objects are defined in space, the function identifies the likelihood that both rainfall systems belong to each other (Fig. 5.2). However, rainfall systems are often characterized by evolution in time as well as spatial structure (Davis et al., 2009b). The integration of the temporal dimension into the object-based analysis marks an important challenge to rainfall systems analysis.

Advances in "big data" analysis have given a new perspective to analyze rainfall systems in space and time. Spatiotemporal object-based methods are currently

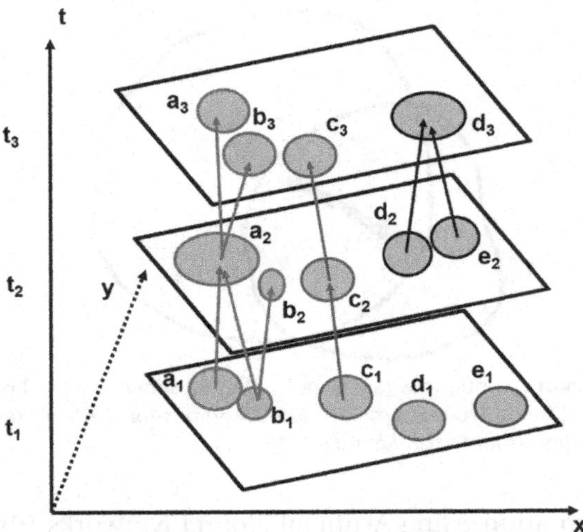

FIGURE 5.2 Storm event tracking based on centroids. The *arrows* represent vector motions of rainfall fields *a*, *b*, *c*, *d*, and e in time steps t_1, t_2, t_3. Colors describe the storm object geometry. *Liu, W., Li, X., Rahn, D.A., 2016. Storm event representation and analysis based on a directed spatiotemporal graph model. International Journal of Geographical Information Science, 30, 948–969.*

developed to identify and compare the temporal and spatial evolution of rainfall systems. This new approach transforms gridded rainfall values into a spatiotemporal dimensional grid (latitude, longitude, time) as volumetric pixels or "voxels." The algorithm identifies objects in space and time connected by a common attribute to obtain a 4D object (latitude, longitude, time, and intensity). Several features can be identified from the type of object such as volume area, duration, average speed, and centroid, among others.

In the literature, there are a couple of examples of this method. For example, Sellars et al. (2013) created PERSIANN-CONNECT, a 4D rainfall dataset, created from the high-resolution PERSIANN satellite-based rainfall data to identify the characteristics of rainfall systems at a large scale. This dataset is stored in a PostgreSQL database and can be easily consulted at http://connect.eng.uci.edu/. Another example developed by this approach is the spatiotemporal verification method MODE Time Domain (TD) (Davis et al., 2009b), an object-based verification method created to evaluate the forecast from high-resolution numerical weather prediction (NWP) models. MODE-TD follows the same methodology of the spatial verification MODE but incorporates temporal evolution in the precipitation system. Mittermaier and Bullock (2013) compared MODE and MODE-TD methods over the United Kingdom for evaluating the spatial and temporal characteristics of cloud cover forecast from km-scale NWP models. This comparison showed notable differences between the two methods. The inclusion of the time dimension provided a different perspective on the verification; however, the choice of method-specific parameters plays a critical role in identifying rainfall objects over time (Fig. 5.3).

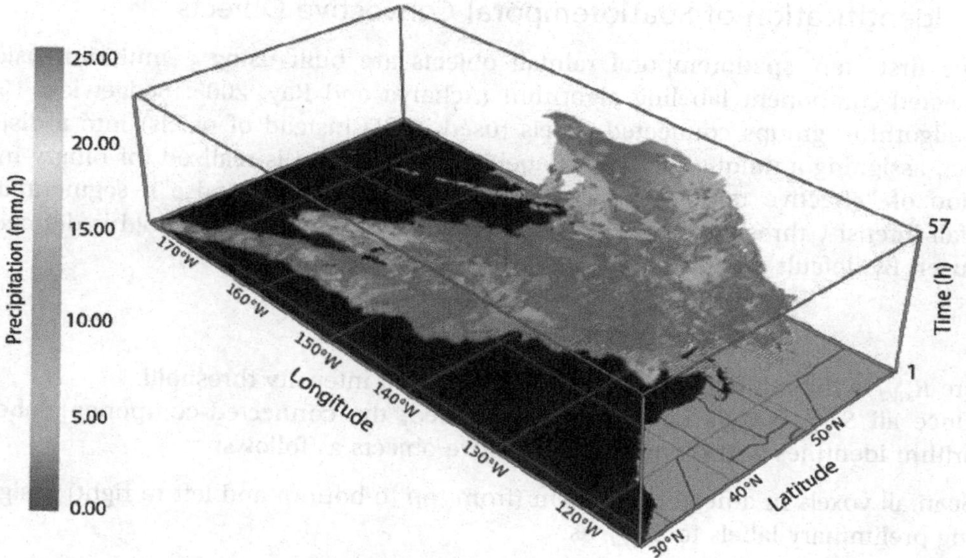

FIGURE 5.3 Example from a 4D rainfall object from PERSIANN-CONNECT. *Sellars, S., Nguyen, P., Chu, W., Gao, X., Hsu, K., Sorooshian, S., 2013. Computational Earth science: big data transformed into insight. Eos Transactions American Geophysical Union, 94, 277–278.*

4. A New Spatiotemporal Object-Based Method for Small-Scale Rainfall Events

Object-based methods in space and time are becoming a powerful tool to analyze the complete structure of large-scale rainfall systems such as events like typhoons, hurricanes, or cold fronts and evaluate their prediction. However, in the case of small-scale (catchment-scale) convective systems, this analysis is challenging (Sillmann et al., 2017). Floods caused by this type of event, e.g., mesoscale and storm-scale rainfall events, produce extreme flash flooding, causing major human and economic damage. Understanding rainfall dynamics plays an important role in hydrological applications. The variation in shape and location of rainfall events affects the runoff volume over the catchment (e.g., Arnaud et al., 2002; Foufoula-Georgiou and Vuruputur, 2001; Haile et al., 2011). To analyze extreme rainfall events at catchment scale, a new object-based method in space and time is proposed. This method enables the feature extraction of different types of rainfall fields from satellite products through a multidimensional connected-component labeling algorithm. This algorithm associates connected rainfall intensities in space and time to create 4D rainfall structures. Afterward, extreme events are selected based on a critical mass threshold. Finally, different types of extreme events are classified based on hydrometeorological criteria. Multiple applications arise from this approach such as the analysis of feature conditions and validation of satellite-based rainfall data.

4.1 Identification of Spatiotemporal Convective Objects

In the first step, spatiotemporal rainfall objects are built using a multidimensional connected-component labeling algorithm (Acharya and Ray, 2005; Sedgewick, 1998). This algorithm groups connected voxels (used in 3D instead of pixels) into a disjoint object, assigning a unique identifier (label). This operation is realized for binary information of "effective rainfall" voxels $S_{[x,y,t]}$ (1 = "true" or 0 = "false"), segmented by rainfall intensity thresholding (Eq. 5.1). The choice of intensity threshold is defined by the user. By default this threshold is 1 mm/h (e.g., Ebert et al., 2009):

$$S_{x,y,t} := \begin{cases} 1, & \text{if } R_{x,y,t} \geq IT, \\ 0, & \text{otherwise} \end{cases} \tag{5.1}$$

where $R_{x,y,t}$ is the rainfall voxel and IT is the rainfall intensity threshold.

Once all $S_{[x,y,t]}$ voxels have been determined, the connected-component labeling algorithm identifies spatiotemporal convective objects as follows:

1. Scan all voxels in a neighbor system (from top to bottom and left to right) assigning preliminary labels to $S_{[x,y,t]}$ as:

$$c(S_{[x,y,t]}) = \{N_{[x,y,t]} \in \partial s \colon S_S = S_N\} \tag{5.2}$$

where $c(S_{[x,y,t]})$ is the preliminary label, S_S,S_N are the properties of the voxel $S_{[x,y,t]}$ and its neighbors $N_{[x,y,t]}$, respectively, and ∂s is the neighbor system in space and time (Fig. 5.4).

2. If a neighbor has more than two $c(S_{[x,y,t]})$, it is assigned the lower label recording the label equivalences in a union-find table.
3. Resolve the table of equivalences classes using the union-find algorithm (Sedgewick, 1998).
4. Make a second iteration relabeling $c(S_{[x,y,t]})$ on the resolved equivalences classes.

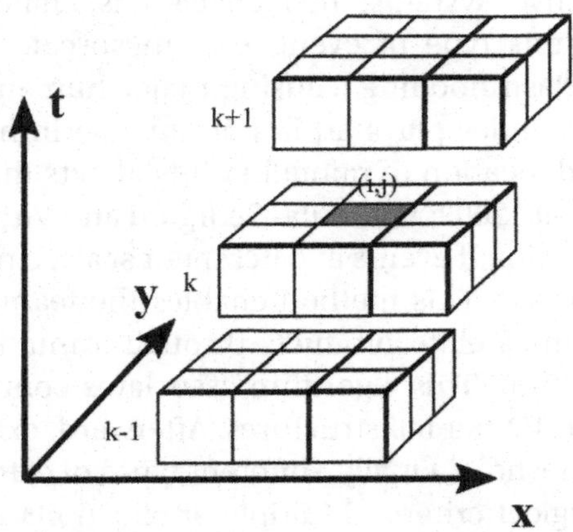

FIGURE 5.4 Neighbor system in space and time (26 voxel neighbors).

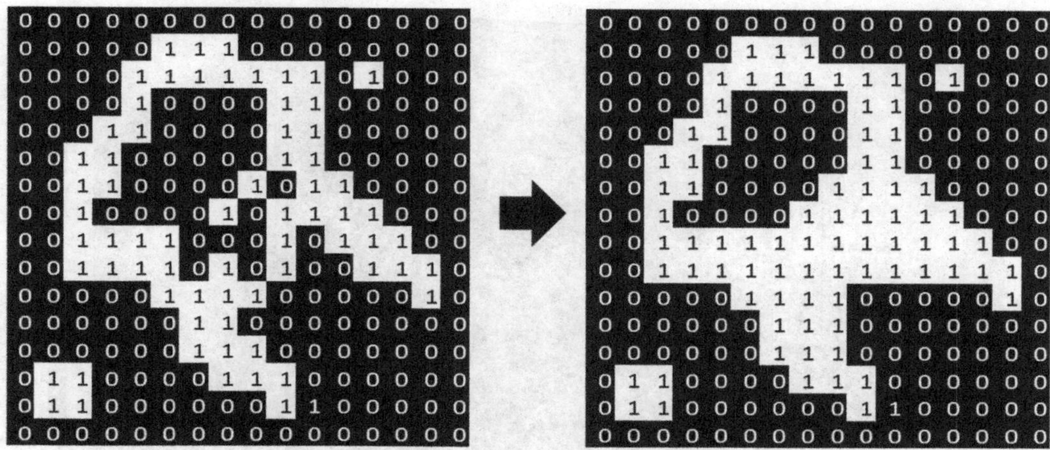

FIGURE 5.5 Morphological closing.

After connected-component labeling is applied, the methodology applies two additional algorithms: a size filtering for noise removal and morphological closing to delineate the 4D objects. A size-filtering algorithm removes objects lower than a size threshold T defined as noise. The selection of T is defined based on the spatial and temporal resolution of the rainfall product (default six voxels). The second algorithm solves false merging resulting from the labeling component algorithm. A morphological closing algorithm divides or merges convective objects with low or strong connectivity (Fig. 5.5). This algorithm uses a dilation erosion process similar to the one employed in object-based algorithms to separate zones with a weak connection in weather radar (e.g., Han et al., 2009). For dilation, boundaries of convective objects are expanded, while for erosion those boundaries are removed. This procedure performs first a morphological dilation followed by an erosion to delineate the convective object.

4.2 Selection of Extreme Rainfall Events

In the second step, convective extreme events are identified based on the critical mass threshold. This parameter corresponds to the minimum volume of rainfall (km^3) necessary to be considered as an extreme event (Grams et al., 2006). The value of the critical mass threshold is defined by the user and is normally obtained depending on the maximum extension and the convective object volume (e.g., Demaria et al., 2011).

Once all rainfall events are identified, a descriptive statistical algorithm calculates diverse characteristics such as the total volume of the storm event (m^3), maximum intensity (mm/h), maximum area (km^2), storm duration (h), and weighted centroid (latitude, longitude, time). This algorithm organizes the event properties in a data structure (N x d dataframe) for easy analysis and visualization (Fig. 5.6).

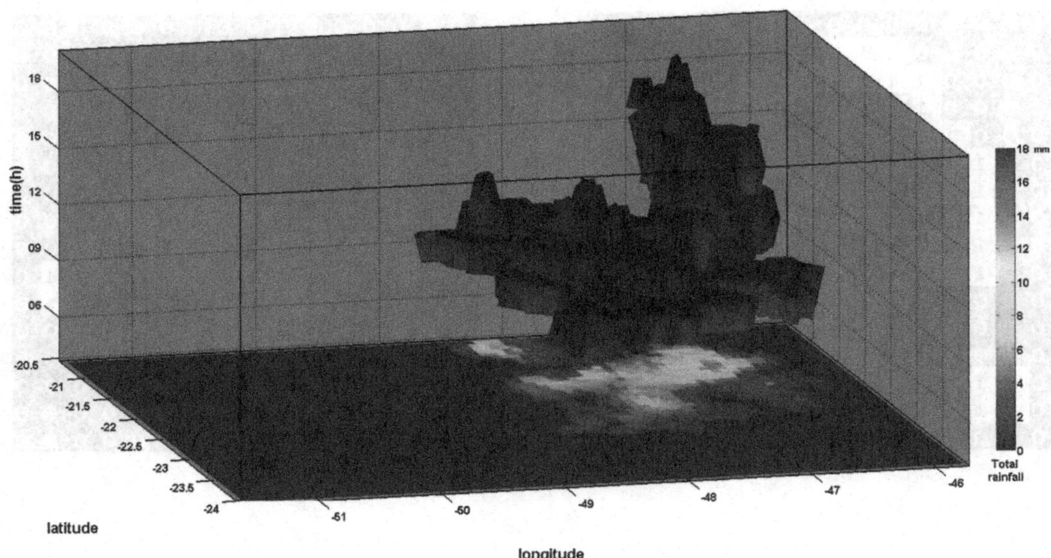

FIGURE 5.6 Example of a rainfall object. Rainfall event over the subtropical catchment of the Tiete River, Brazil (February 24, 2008).

4.3 Classification of Extreme Event

The last step corresponds to the classification of extreme rainfall events by hydrometeorological criteria. Based on three main characteristics of extreme event (maximum area, duration, and volume of rainfall) (Fig. 5.7), four types of rainfall events are identified: local and short extreme events (LSE); local and long-duration extreme events (LLE); spatially extensive extreme events (SEE); and long-duration and spatially extensive extreme events (SLE). LSE and LLE are associated with small convective systems with a slow or fast movement, which contain a large amount of rainfall falling over a reduced area (e.g., a city, small draining catchment), for example, convective storms and rainfall cells, among others. SEE and LSE are events with a slow or fast motion, which are extended over large areas (e.g., mesoscale level).

4.4 Spatiotemporal Verification

Several error metrics can be applied to rainfall objects to evaluate the performance of satellite products starting from standard verification methods such as continuous statistics (e.g., mean error, RMSE, correlation coefficient) and categorical verifications (e.g., BIAS, POD, FAR) to most sophisticated diagnostics verification methods such as error decomposition (Ebert and McBride, 2000), correspondence ratio (Stensrud and Wandishin, 2000), or displacement and amplitude score (Keil and Craig, 2009). This method can be used for identification at different resolutions and scales.

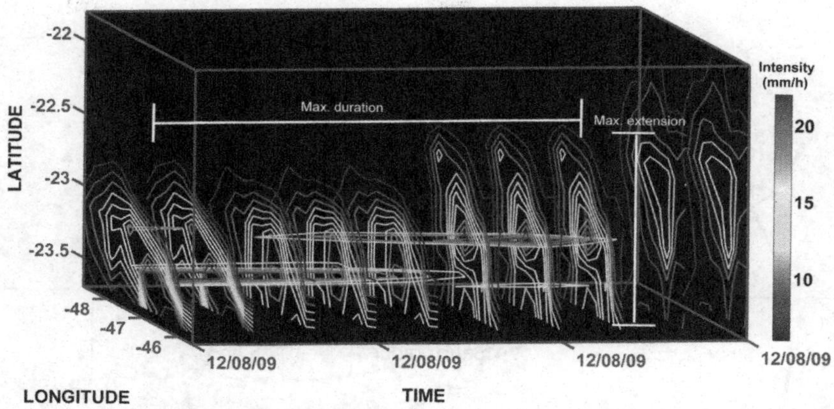

FIGURE 5.7 Spatiotemporal characteristics of extreme rainfall.

5. Applications Over a Subtropical Catchment in Southeastern Brazil

5.1 Study Area and Datasets

The study site for this research is the subtropical catchment of the Tiete River (Fig. 5.8). The area is part of the Parana River basin, which is one of the main river systems in Brazil. With a drainage area of 150,000 km^2, the catchment is densely populated and contains several of the state's large cities. Due to the location, the area is strongly impacted by inter- and extratropical climatic conditions, e.g., the South American monsoon system (Boers et al., 2013). This factor makes the area prone to extreme landslides and flash floods (Sprissler, 2011). This research analyzes the climatic conditions during the core of the monsoon in South America from December to February from 2007 to 2017.

NRT satellite rainfall information from CMORPH (Joyce et al., 2004) is used for the study. The selection of this dataset responds to a previous study in the area made by Laverde-Barajas et al. (n.d.), where this dataset showed a superior performance in estimating different types of rainfall events. CMORPH is used at 8 km (0.0727 degrees) with half hour temporal resolution (1 h aggregated). Ground measurement is obtained by the weather radar station Bauru (CAPPI 3.5) from the Faculty of Sciences of the Sao Paulo State University. Weather rainfall data were bias corrected using gauge stations from automatic weather stations obtained by the *Centro Integrado de Informações Agrometeorológicas* (CIIAGRO, 2017). Ground measurements used have a spatial resolution of 1k (0.01 degrees) with every 15 min (aggregated to 1 h).

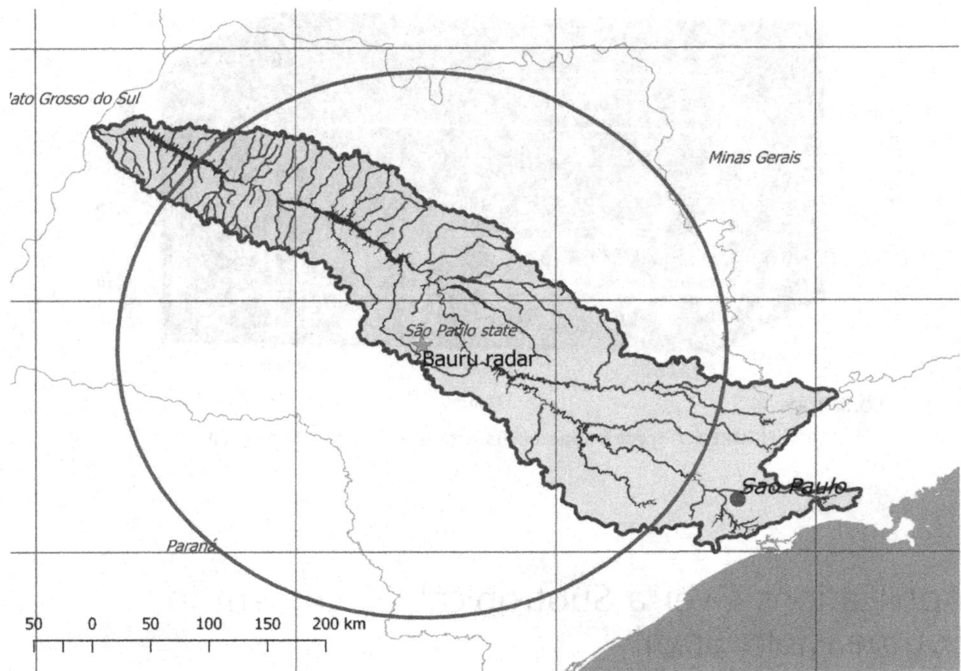

FIGURE 5.8 Catchment of the Tiete River. *Red circle* is the domain of the weather radar.

5.2 Storm Identification

The application of this methodology requires the definition of two parameters: the convective threshold and the critical mass threshold. The definition of convective threshold is critical in the definition of rainfall systems. Several studies in spatial verification have found that reduction of this threshold allows the incorporation of small systems with the deficiency of creating unrealistic rainfall objects (Demaria et al., 2011). In this study area, it was observed that the parameter of the best fit is 1 mm/h. The critical mass threshold value is defined by the relationship between maximum area and volume of rainfall systems. In the Tiete River catchment, it is found that convective systems larger than 5000 km^2 (8% of the population) contribute almost 90% of the rainfall during the monsoon season (Table 5.1). This area corresponds to a total volume of around 0.14 km^3 (Fig. 5.9). Based on this result, a critical mass threshold of 0.1 km^3 was selected.

By the application of this method over hourly information from CMORPH in the catchment area, 694 extreme rainfall events were identified during the monsoon season from 2007 to 2017. For each rainfall event, several descripted characteristics were extracted. These include volume, intensity, maximum area, duration, and weighted centroid.

Table 5.1 Contribution of the Rainfall Fields Into Total Amount of Water During the Monsoon Season

Rainfall Features	Population Fraction (%)	Rainfall Contribution (%)
Size range 75–500 km^2	1	55
Size range 500–2,000 km^2	5	28
Size range 2,000–5,000 km^2	8	8
Size range 5,000–10,000 km^3	11	3
More than 10,000 km^2	76	5

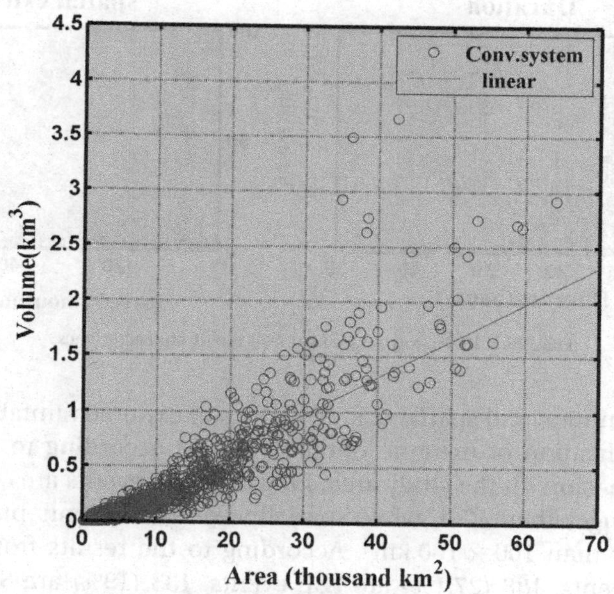

FIGURE 5.9 Relationship between maximum event area and total rainfall volume.

5.3 Spatiotemporal Characteristics of Extreme Rainfall Events

Fig. 5.10 presents the histograms of extreme rainfall systems in terms of total volume, maximum rainfall, duration, and maximum extension (area). According to the results, the total volume of the rainfall presented a positively skewed distribution commonly associated with these systems. On the other hand, maximum rainfall values of the events followed a normal distribution ranging from 5 to 50 mm/h with a median value of 30 mm/h. In relation to spatiotemporal characteristics, rainfall events are frequently longer in time (mean = 14 h STD = 6 h) normally varying between 8 and 18 h. In space, the maximum extension of the events ranges between 6000 and 12,000 km^2 covering between 4% and 8% of the total catchment area.

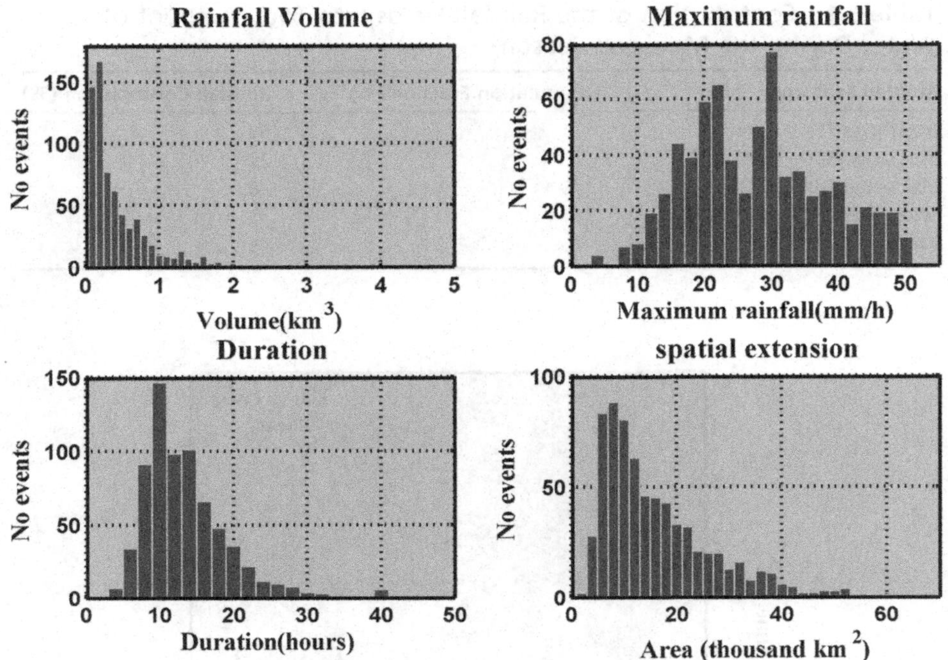

FIGURE 5.10 Histograms of rainfall event characteristics.

Based on the temporal and spatial characteristics of extreme rainfall systems, Fig. 5.11 presents the classification of extreme rainfall systems according to volume, duration, and maximum extension. In the study area, long duration events are considered systems with a duration longer than 12 h, while spatially extensive events present a maximum extension of larger than $100 \times 100\ \text{km}^2$. According to the results from 694 events, 314 (45.3%) are SLE events, 188 (27.1%) are LSE events, 133 (19%) are SEE events, and 59 (8.4%) are LLE events.

5.4 Application for Satellite-Based Rainfall Products Verification

The capacities of this methodology for product verification are evaluated for two types of extreme rainfall systems. The first event is an LSE event that occurred on February 12, 2011 between 16:00 and 21:00 h causing several flash floods in the southeastern part of the Tiete River (SIMPAT). The second event is an SLE event that took place over the region between January 11 and 12, 2011. This rainfall event was one of the most important extreme events in the area, where the extreme rainfall triggered massive floods and landslides over important cities of the state. Thirteen people lost their lives and thousands lost their homes and other buildings (NOAA, 2011). NRT satellite-based information from CMORPH is compared against weather radar located in the study area (Fig. 5.8). Both products are compared based on their main characteristics

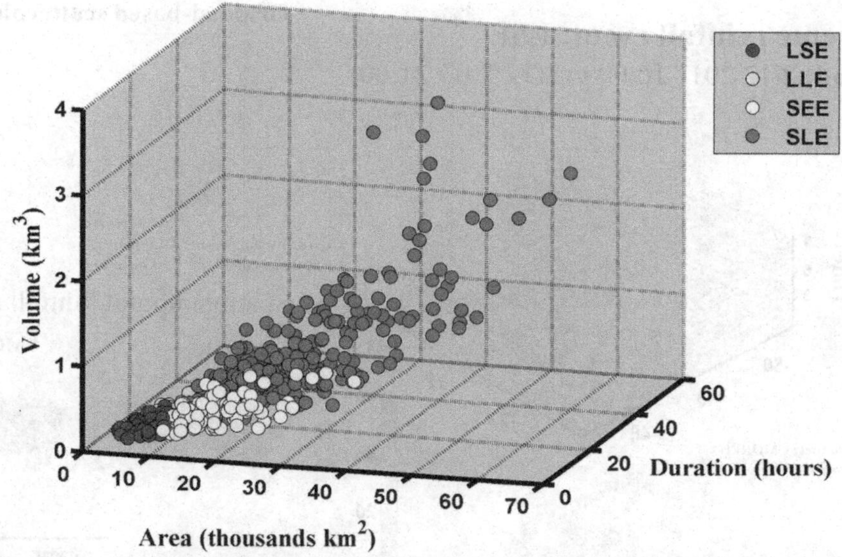

FIGURE 5.11 Classification of extreme rainfall events. *LLE*, Local and long-duration extreme events; *LSE*, local and short extreme events; *SEE*, spatially extensive extreme events; *SLE*, long-duration and spatially extensive extreme events.

(maximum, volume, duration, and area) and their performance was evaluated grid to grid in terms of the level of displacement (with respect to the weighted centroid), RMSE, and correlation coefficient. In this latter case, weather radar was upscaled to match the grid size of the satellite product.

Fig. 5.12 displays the verification used for the LSE event of February 12, 2011. The results showed a high rainfall concentration over the southeastern part of the region in both products. CMORPH product and the weather radar objects presented similar structures in terms of shape but had important differences in terms of magnitude. CMORPH tended to underestimate the magnitude of the extreme event and the total volume of the system. The characteristics in space and time presented a slight displacement to the northwest in space and a displacement in time of 1 h. Radar and CMORPH products covered similar extensions (8974 and 8036 km^2); however, CMORPH had a short duration (7 and 6 h). Based on two continuous verification metrics, CMORPH had a maximum correlation (shifted) of 0.16 and RMSE of 5.5 mm/h.

Verification of the SLE event on January 11, 2011 is presented in Fig. 5.13. According to the results, CMORPH was able to capture important characteristics of the extreme events. In the case of magnitude, both 4D objects presented similar characteristics. Maximum rainfall value was slightly overestimated (~5 mm/h), as was volume (~0.4 km^3). In the case of space and time characteristics, CMORPH presented a considerable shifting to the northwestern part of the catchment. The maximum extension estimated covers 403,200 km^2 being greater than the radar estimate (323,500 km^2).

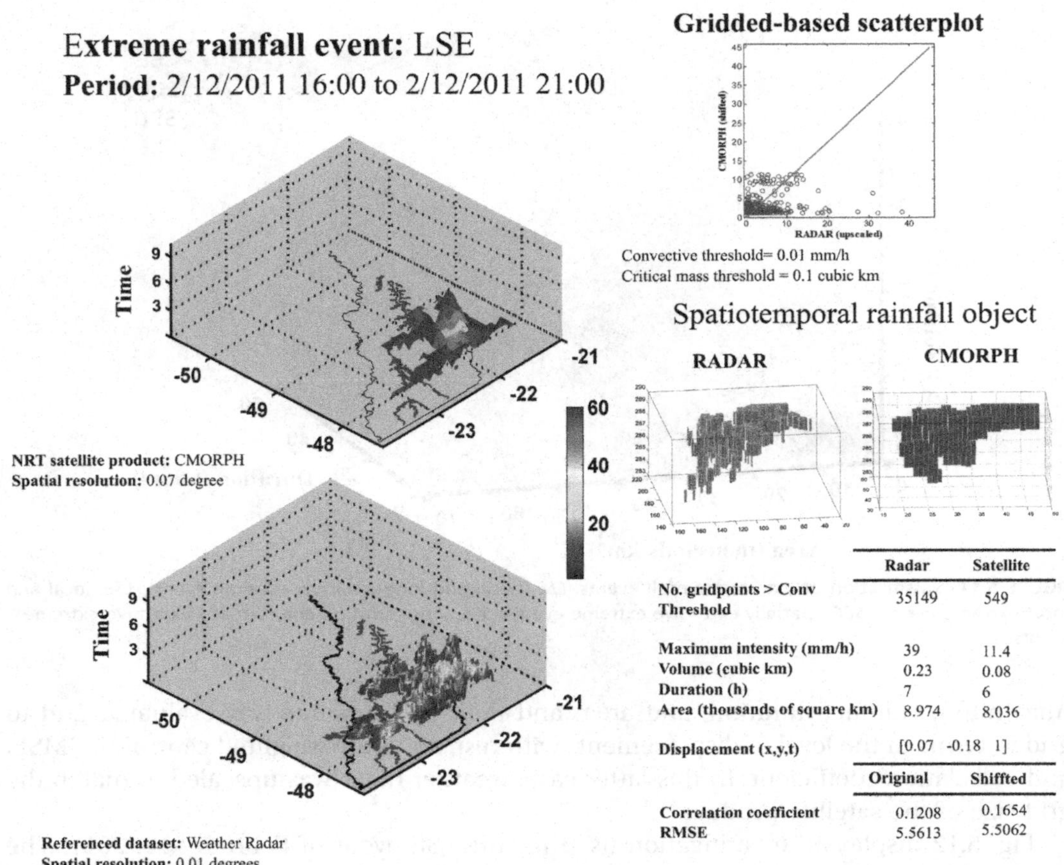

FIGURE 5.12 Spatiotemporal verification for local and short extreme rainfall events. *CMORPH*, Climate prediction center morphing technique; *NRT*, near-real time.

In both cases, the rainfall event had a duration of 16 h over the area. Evaluation in a gridded-based approach displayed a maximum correlation of 0.175 and an RMSE of 7.36 mm/h.

6. Final Remarks

Object-based methods offer a unique perspective to analyze rainfall systems, providing a complete picture of the dynamics of rainfall fields. In this research, the analysis of rainfall events at catchment scale goes a step further analyzing extreme events in space and time. This method provides a complete diagnosis of extreme events identifying a large number of characteristics such as volume, area, duration, orientation, and speed,

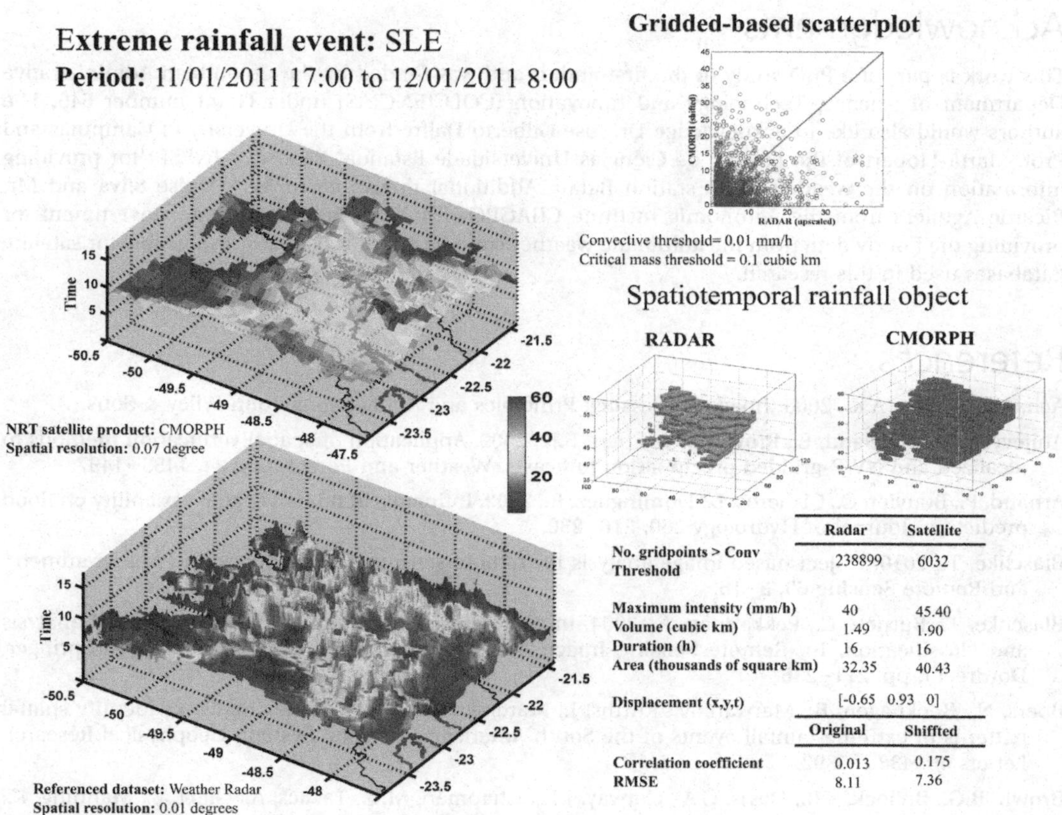

FIGURE 5.13 Spatiotemporal verification for long-duration and spatially extensive extreme rainfall events. *CMORPH*, Climate prediction center morphing technique; *NRT*, near-real time.

among others. The proposed methodology is very flexible and could be applied to different regions through the adjustment of two parameters: the convective threshold and the critical mass threshold. These parameters are especially sensitive and they can be fundamental in the analysis.

This method opens the door to new discoveries in hydrological sciences toward understanding rainfall dynamics over catchments in extreme conditions. Possible new developments could lead to a better understanding of the flood generation process evaluating the relationship between rainfall pattern and runoff. This can induce new ways of describing flood events based on the spatiotemporal characteristics of rainfall events. Additionally, other remote sensing products may be evaluated using this approach, for instance, the recently released Global Precipitation Measurement satellite product from NASA, among others. This method could be used to evaluate the capabilities of high-resolution rainfall products over a determined area.

Acknowledgments

This work is part of a PhD study of the first author and was funded by the Colombian Administrative Department of Science, Technology and Innovation (COLCIENCIAS) under Grant number 646. The authors would also like to acknowledge Dr. Jose Gilberto Dalfre from the University of Campinas and Prof. Marta Llopart of the Faculty de Ciências Universidade Estadual Paulista (UNESP) for providing information on the weather radar station Barau. Additional thanks go to Ms. Denise Silva and Mr. Ricardo Aguilera from the Agronomic Institute CIIAGRO–FUNDAG São Paulo State Government for providing the hourly data from the automatic weather stations and the agencies responsible for satellite databases used in this research.

References

Acharya, T., Ray, A.K., 2005. Image Processing: Principles and Applications. John Wiley & Sons.

Ahijevych, D., Gilleland, E., Brown, B.G., Ebert, E.E., 2009. Application of spatial verification methods to idealized and NWP-gridded precipitation forecasts. Weather and Forecasting 24, 1485–1497.

Arnaud, P., Bouvier, C., Cisneros, L., Dominguez, R., 2002. Influence of rainfall spatial variability on flood prediction. Journal of Hydrology 260, 216–230.

Blaschke, T., 2010. Object based image analysis for remote sensing. ISPRS Journal of Photogrammetry and Remote Sensing 65, 2–16.

Blaschke, T., Burnett, C., Pekkarinen, A., 2004. Image segmentation methods for object-based analysis and classification. In: Remote Sensing Image Analysis: Including the Spatial Domain. Springer, Dordrecht, pp. 211–236.

Boers, N., Bookhagen, B., Marwan, N., Kurths, J., Marengo, J., 2013. Complex networks identify spatial patterns of extreme rainfall events of the South American Monsoon System. Geophysical Research Letters 40, 4386–4392.

Brown, B.G., Bullock, Y.R., Davis, C.A., Gotway, J.H., Chapman, M.B., Takacs, A., Gillel, E., Manning, K., Mahoney, J.L., 2004. New verification approaches for convective weather forecasts. In: Preprints, 11th Conference on Aviation, Range, and Aerospace, Hyannis, pp. 3–8.

CIIAGRO, C.I. de I.A., 2017. Rede Meteorológica Automática.

Davis, C.A., Brown, B.G., Bullock, R., Halley-Gotway, J., 2009a. The method for object-based diagnostic evaluation (MODE) applied to numerical forecasts from the 2005 NSSL/SPC spring program. Weather and Forecasting 24, 1252–1267.

Davis, C.A., Brown, B., Bullock, R., 2009b. Spatial and temporal object-based evaluation of numerical precipitation forecasts. In: Preprints, 23rd Conf. on Weather Analysis and Forecasting/19th Conf. on Numerical Weather Prediction, Omaha, NE, Amer. Meteor. Soc. A.

Demaria, E.M.C., Rodriguez, D.A., Ebert, E.E., Salio, P., Su, F., Valdes, J.B., 2011. Evaluation of mesoscale convective systems in South America using multiple satellite products and an object-based approach. Journal of Geophysical Research 116, D08103.

Desclée, B., Bogaert, P., Defourny, P., 2006. Forest change detection by statistical object-based method. Remote Sensing of Environment 102, 1–11.

Dixon, M., Wiener, G., 1993. TITAN: thunderstorm identification, tracking, analysis, and nowcasting—a radar-based methodology. Journal of Atmospheric and Oceanic Technology 10, 785–797.

Ebert, E.E., 2007. Methods for verifying satellite precipitation estimates. In: Measuring Precipitation from Space. Springer, pp. 345–356.

Ebert, E.E., McBride, J.L., 2000. Verification of precipitation in weather systems: determination of systematic errors. Journal of Hydrology 239, 179–202.

Ebert, E.E., William, A., Gallus, Jr., 2009. Toward better understanding of the contiguous rain area (CRA) method for spatial forecast verification. Weather and Forecasting 24.5, 1401–1415.

Foufoula-Georgiou, E., Vuruputur, V., 2001. Patterns and Organization in Precipitation. Cambridge University Press.

Gilleland, E., Ahijevych, D., Brown, B.G., Casati, B., Ebert, E.E., 2009. Intercomparison of spatial forecast verification methods. Weather and Forecasting 24, 1416–1430.

Grams, J.S., Gallus, W.A., Koch, S.E., Wharton, L.S., Loughe, A., Ebert, E.E., 2006. The use of a modified bbert–McBride technique to evaluate mesoscale model QPF as a function of convective system morphology during IHOP 2002. Weather and Forecasting 21, 288–306.

Grayson, R., Blöschl, G., 2001. Spatial Patterns in Catchment Hydrology: Observations and Modelling (CUP Archive).

Haile, A., Rientjes, T., Habib, E., Jetten, V., Gebremichael, M., 2011. Rain Event Properties at the Source of the Blue Nile River. Copernic. Publ.

Han, L., Fu, S., Zhao, L., Zheng, Y., Wang, H., Lin, Y., 2009. 3D convective storm identification, tracking, and forecasting—an enhanced TITAN algorithm. Journal of Atmospheric and Oceanic Technology 26, 719–732.

Huffman, G.J., Bolvin, D.T., Nelkin, E.J., Wolff, D.B., Adler, R.F., Gu, G., Hong, Y., Bowman, K.P., Stocker, E.F., 2007. The TRMM multisatellite precipitation analysis (TMPA): quasi-global, multiyear, combined-sensor precipitation estimates at fine scales. Journal of Hydrometeorology 8, 38–55.

Johnson, J.T., MacKeen, P.L., Witt, A., Mitchell, E.D.W., Stumpf, G.J., Eilts, M.D., Thomas, K.W., 1998. The storm cell identification and tracking algorithm: an enhanced WSR-88D algorithm. Weather and Forecasting 13, 263–276.

Joyce, R.J., Janowiak, J.E., Arkin, P.A., Xie, P., 2004. CMORPH: a method that produces global precipitation estimates from passive microwave and infrared data at high spatial and temporal resolution. Journal of Hydrometeorology 5, 487–503.

Keil, C., Craig, G.C., 2009. A displacement and amplitude score employing an optical flow technique. Weather and Forecasting 24, 1297–1308.

Knight, L., 2011. World Disasters Report 2011: Focus on Hunger and Malnutrition. International Federation of Red Cross and Red Crescent Societies, Switzerland.

Laverde-Barajas, M., Corzo Perez, G.A., Dalfré Filho, J.G., Solomatine, D., 2018. Assessing the performance of near real-time rainfall products to represent spatiotemporal characteristics of extreme events: case study of a subtropical catchment in south-eastern. Brazil International Journal of Remote Sensing 1–19.

Li, J., Hsu, K., AghaKouchak, A., Sorooshian, S., 2015. An object-based approach for verification of precipitation estimation. International Journal of Remote Sensing 36, 513–529.

Li, J., Hsu, K.-L., AghaKouchak, A., Sorooshian, S., 2016. Object-based assessment of satellite precipitation products. Remote Sensing 8, 547.

Liu, W., Li, X., Rahn, D.A., 2016. Storm event representation and analysis based on a directed spatiotemporal graph model. International Journal of Geographical Information Science 30, 948–969.

Mittermaier, M.P., Bullock, R., 2013. Using MODE to explore the spatial and temporal characteristics of cloud cover forecasts from high-resolution NWP models. Meteorological Applications 20, 187–196.

NOAA, N.O. and A.A, 2011. Global Hazards - January 2011.

Sedgewick, R., 1998. Algorithms in C, Parts 1–4: Fundamentals, Data Structures, Sorting, Searching. Addison-Wesley Professional, Reading, Mass.

Sellars, S., Nguyen, P., Chu, W., Gao, X., Hsu, K., Sorooshian, S., 2013. Computational Earth science: big data transformed into insight. Eos Transactions American Geophysical Union 94, 277–278.

Sillmann, J., Thorarinsdottir, T., Keenlyside, N., Schaller, N., Alexander, L.V., Hegerl, G., Seneviratne, S.I., Vautard, R., Zhang, X., Zwiers, F.W., 2017. Understanding, modeling and predicting weather and climate extremes: challenges and opportunities. Weather and Climate Extremes 18, 65–74.

SIMPAT, S.I. de M., Previsão e Alerta de Tempestades para as Regiões Sul-Sudeste do Brasil banco de Dados de desastres naturais.

Skok, G., Tribbia, J., Rakovec, J., Brown, B., 2009. Object-based analysis of satellite-derived precipitation systems over the low- and midlatitude Pacific Ocean. Monthly Weather Review 137, 3196–3218.

Sorooshian, S., Hsu, K.-L., Gao, X., Gupta, H.V., Imam, B., Braithwaite, D., 2000. Evaluation of PERSIANN system satellite-based estimates of tropical rainfall. Bulletin of the American Meteorological Society 81, 2035–2046.

Sprissler, T., 2011. Flood Risk Brazil: Prevention, Adaptation and Insurance (Switzerland).

Stensrud, D.J., Wandishin, M.S., 2000. The correspondence ratio in forecast evaluation. Weather and Forecasting 15, 593–602.

Vörösmarty, C.J., Guenni, L.B. de, Wollheim, W.M., Pellerin, B., Bjerklie, D., Cardoso, M., D'Almeida, C., Green, P., Colon, L., 2013. Extreme rainfall, vulnerability and risk: a continental-scale assessment for South America. Philosophical Transactions of the Royal Society A 371, 20120408.

Wernli, H., Paulat, M., Hagen, M., Frei, C., 2008. SAL—a novel quality measure for the verification of quantitative precipitation forecasts. Monthly Weather Review 136, 4470–4487.

Wilhelmi, O.V., Morss, R.E., 2013. Integrated analysis of societal vulnerability in an extreme precipitation event: a Fort Collins case study. Environmental Science & Policy 26, 49–62.

Yu, B., Liu, H., Wu, J., Hu, Y., Zhang, L., 2010. Automated derivation of urban building density information using airborne LiDAR data and object-based method. Landscape and Urban Planning 98, 210–219.

6

Spatial and Temporal Variations' of Habitat Suitability for Fish: A Case Study in Abras de Mantequilla Wetland, Ecuador

Gabriela Alvarez-Mieles[1,2,3], Gerald Corzo[1], Arthur E. Mynett[1,2]
[1]*IHE DELFT INSTITUTE OF WATER EDUCATION, DELFT, THE NETHERLANDS;*
[2]*DELFT UNIVERSITY OF TECHNOLOGY, FACULTY CITG, DELFT, THE NETHERLANDS;*
[3]*UNIVERSIDAD DE GUAYAQUIL, FACULTAD DE CIENCIAS NATURALES, GUAYAQUIL, ECUADOR*

1. Introduction

1.1 Wetlands and Tropical Floodplains

Wetlands are among the most productive environments in the world. They are crucial for the maintenance of biological biodiversity, providing water and primary production upon which numerous species of fauna and flora depend for survival (Halls, 1997; Ramsar, 2013). Tropical rivers associated with floodplain areas provide dynamic habitats for fish (Winemiller and Jepsen, 1998), and contribute to maintaining the biodiversity of the whole river ecosystem, provided that connectivity is maintained. Connectivity is a key issue in floodplains, since richness and diversity of species decrease with decreasing hydrological connectivity (Aarts et al., 2004). Tropical floodplains faced gradual drying due to anthropogenic activities such as dams and irrigation, causing impacts on fish communities. A reduction in the inundated areas of floodplains decreases the habitat availability for fish communities. Reduction in habitat areas in turn produces an increase in fish densities (per unit surface area), intensification in species interaction, and competition for resources (Winemiller and Jepsen, 1998). In South America, designation of aquatic protected areas has now started, and fish studies have been focused more frequently on local rather than on river basin scales (Barletta et al., 2010).

Spatiotemporal Analysis of Extreme Hydrological Events. https://doi.org/10.1016/B978-0-12-811689-0.00006-9

1.2 Habitat Analysis and History

A series of Habitat Suitability Index (HSI) models was developed in the early 1980s to provide habitat information of several wildlife species in the United States (Schamberger et al., 1982), although initial habitat studies were started in the 1950s to determine suitable areas for salmon spawning (Jowett, 1997). Habitat suitability models are expressed with a numerical index on a 0.0–1.0 scale, with the assumption that there is a positive relationship between the index and the habitat carrying capacity of the selected species (Schamberger et al., 1982). Habitat suitability criteria can be expressed in different categories and formats (Bovee, 1982). A first category is based on expert opinions (professional, stakeholders) instead of data. A second category is based on data where organisms of the target species were collected. Thus they are known as "utilization or habitat use functions" because they represent the conditions that the target species faced at the time of observation or sampling. However, these criteria can be biased by the environmental conditions available at the observation time, since organisms could be forced to use suboptimal conditions when optimal conditions are not available. To correct this function bias and be less site specific, a third category "preference functions" can be created (Bovee, 1986; Bovee et al., 1998).

Habitat methods can be considered an expansion of hydraulic methods, where the evaluation of flow requirements is based on hydraulic conditions that aim to meet biological requirements. Hydraulic model outputs such as water depth and velocity are subsequently evaluated with habitat suitability criteria for a specific species or group of species. In use since 1970s, PHABSIM, a physical habitat model, is a component of the Instream Flow Incremental Methodology, a decision-support system designed to help managers in the evaluation of different water management alternatives (Bovee et al., 1998). This model was developed by the US Fish and Wildlife Service to analyze the relationships between flow and physical habitat (Milhous and Waddle, 2012; Spence and Hickley, 2000). The development of fish habitat models (CASIMIR) started in the early 1990s at the University of Stuttgart to investigate the impacts related to hydropower operations (Kerle et al., 2002; Schneider et al., 2010). Habitat models such as PHABSIM and CASIMIR use expert knowledge or statistical analysis of field data to describe biotic/abiotic relationships (Tuhtan and Wieprecht, 2012). Habitat suitability is then evaluated by the development and further application of membership functions, also called "preference functions". Habitat suitability models have been widely applied to fish in temperate areas (Brown et al., 2000; Costa et al., 2012; Kerle et al., 2002; Mouton et al., 2011; Muñoz-Mas et al., 2012; Schneider et al., 2010, 2012; Tuhtan and Wieprecht, 2012; Zorn et al., 2012).

In the Netherlands, a framework for habitat modeling based on an ecosystem approach started in the 1990s (Duel et al., 2003). This approach includes four steps: first, the spatial distribution of ecotopes is simulated. For this step hydro- and morphodynamics (stream velocity, depth, flood frequency), drivers that define the river ecotopes, are evaluated. Steps two and three assess the availability and suitability of these habitats for selected species. For habitat suitability, requirements (shelter, food, nutrients) and threats (pollutants, toxic chemicals, etc.) for target species are considered. These habitat requirements are usually derived from field observations, historical data, and statistical analysis of the environmental factors that belong to the habitats where the target species live. From this analysis, habitat suitability is determined by the environmental factors that limit its carrying capacity. A fourth step evaluates the connectivity of suitable habitats into ecological networks (Duel et al., 2003). HABITAT, a spatial tool for ecological assessment, has been applied in the Netherlands as a decision support system for implementation of the Birds, Habitats, and Water Framework Directive (Haasnoot, 2009a). Membership functions for several temperate fish species are available in the HABITAT toolbox (DELTARES, 2016), and fish studies related to spawning habitat availability have been developed (Wolfshaar et al., 2010). Nevertheless, in tropical systems, a few studies that identify habitat preferences and develop habitat suitability criteria for fish are available (Costa et al., 2013; Teresa and Casatti, 2013). These studies developed habitat suitability criteria in the form of "preference curves" for several fish species based on hydraulic features (depth, velocity) and substrate.

1.3 Habitat and Hydrology

The development of habitat suitability analysis has to be related to the fact that species are distributed according to their preferences for feeding and reproduction (Teresa and Casatti, 2013). Linking target species with their physical, chemical, and biotic conditions is the base of habitat assessment. Physical aspects include hydrology and geomorphology, and hydrological indicators can explain physical, chemical, and biological processes in wetlands (Funk et al., 2013). Thus hydrologic conditions are key drivers for the wetland's structure and function. Hydrology influences several abiotic factors that determine which biota will develop in the wetland. "Hydrology is probably the single most important determinant of the establishment and maintenance of specific types of wetlands and wetland processes" (Mitsch and Gosselink, 2007). Specific mesohabitat characteristics such as depth and velocity have been found to play a key role in explaining fish community structures (Arrington and Winemiller, 2006), with several studies considered as the main variables for fish habitat analysis (Freeman et al., 2001; Schneider et al., 2012; Teresa and Casatti, 2013; Wolfshaar et al., 2010).

1.4 Fish and Relations With Shallow Areas and Macrophytes

Several studies acknowledged the importance of littoral areas as habitats for fish communities (Arrington and Winemiller, 2006; Teixeira-de Mello et al., 2009), and the association of fish to macrophytes (Agostinho et al., 2007; Meerhoff et al., 2007a; Meschiatti et al., 2000). Shallow areas of Abras de Mantequilla wetland (Ecuador) are principally populated by small-sized fish from the Characidae family (Alvarez-Mieles et al., 2013). Characids are an important source of food for higher trophic levels (top fish predators that have a value for local communities) and important seed dispersers in neotropical floodplains. Previous studies in the wetland and associated basin, "Guayas River basin" reported the presence of this family (Florencio, 1993; INP, 2012; Laaz et al., 2009; Prado, 2009; Prado et al., 2012). Some species of this family are common in all the western basins of Ecuador (Gery, 1977; Glodek, 1978; Laaz et al., 2009; Loh et al., 2014), but others are endemic in the "Guayas basin" (Laaz and Torres, 2014; Roberts, 1973). However, information on the ecology or evolutionary history of most fish species in the region is very limited and even lacking (Aguirre et al., 2013). Littoral fish assemblages in Abras de Mantequilla wetland included both common and endemic species. At middle and low wetland areas endemic species such as *Phenacobrycon henni*, *Landonia latidens*, *Iotabrycon praecox*, and *Hyphessobrycon ecuadoriensis* have been collected, thus the importance of assessing the habitat conditions in this tropical wetland. Furthermore, the wetland provides a habitat for fish of commercial interest for local communities: *Aequidens rivulatus*, *Cichlasoma festae*, *Curimatorbis boulengeri*, *Brycon dentex*, and *Ichthyoelephas humeralis*. These species have been collected in the main channels of this wetland, and freely move in the pelagic areas, but they also utilize littoral vegetated areas to protect their eggs after spawning (Barnhill and Lopez, 1974; Florencio, 1993; Quevedo, 2008; Revelo, 2010).

1.5 Our Scope

In this chapter, the following key research questions are investigated:

- Are hydrological conditions an important factor shaping habitat analysis?
- Is there an optimal time period during the year that provides a higher extension of suitable areas?
- Are there specific regions in the wetland more suitable than others?

This study proposes a methodology to quantify the extension of suitable habitat areas for the fish communities of Abras de Mantequilla wetland based on hydrodynamic features. A measure related to the percentage of suitable habitat areas (PSA) is proposed as a tool to explore the temporal and spatial variability of the habitat through the year for different hydrological conditions. We suggest that by protecting the habitat of the fish communities, lower trophic community levels (e.g., macroinvertebrates and plankton) that also inhabit the wetland areas would be also protected.

2. Methodology

2.1 Study Area

The Abras de Mantequilla wetland is located at the center of the Guayas River basin in the coastal region of Ecuador (Fig. 6.1). The wetland was declared a RAMSAR site in 2000 due to the important role in conservation of bird fauna biodiversity, and especially because it supports three migratory species of birds (Ramsar, 2014). It is also an

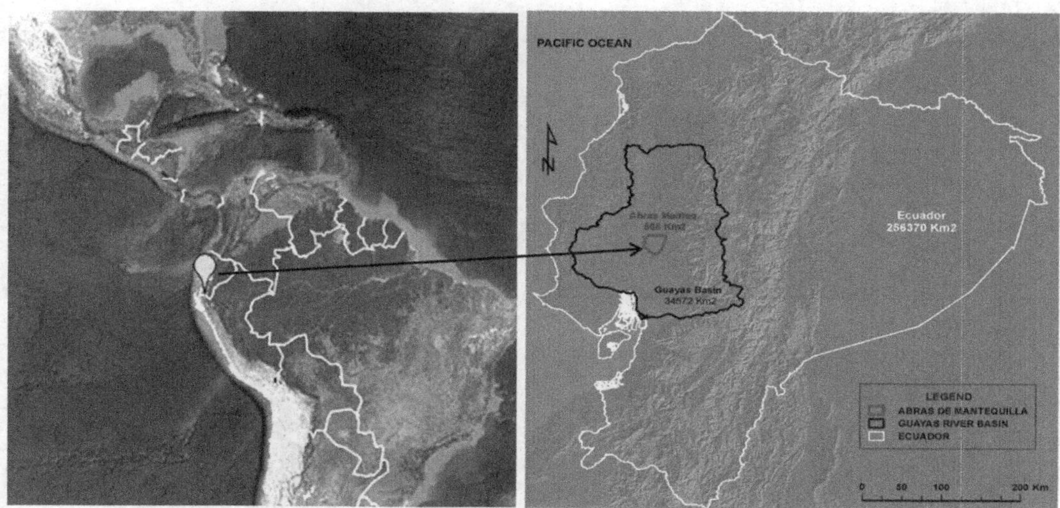

FIGURE 6.1 Study area: Abras de Mantequilla wetland location.

Important Bird and Biodiversity Area (IBA) with 127 bird species reported. The wetland was selected as the South America case study for the WETwin Project, a project funded by the European Commission (FP7) to enhance the role of the wetlands in integrated water resource management. The project included seven study areas in three continents: Europe, Africa, and South America. The main characteristic of these areas is that all of them are inland wetlands related to a river basin. The wetland, part of the Chojampe subbasin consists of branching water courses surrounded by elevations of 5–10 m (Quevedo, 2008). Due to land conversion to agriculture in the last few decades, original forest coverage around the wetland is less than 3%. Agriculture in the surrounding wetland area mainly consists of short-term crops (rice, maize). Current land uses around the wetland and hydropower projects in the upper catchment area are expected to be the main constraints for the future health of the wetland. Littoral areas of the wetland are covered by banks of macrophytes populated by small fish (Characidae family) (Fig. 6.2D). Collected species from this family include common ones such as *Astyanax festae,* and endemic species such as *L. latidens* and *H. ecuadoriensis* (Fig. 6.3). The wetland also provides a habitat for several fish species of commercial interest for local communities (Barnhill and Lopez, 1974; Florencio, 1993; Quevedo, 2008; Revelo, 2010).

Due to the proximity to Equator, there are only two climatic periods: the wet season (mid-December up to mid-May) and the dry season (July–November) (Fig. 6.4). Annual variability in precipitation is depicted in Fig. 6.5. The highest annual precipitation in Pichilingue station was observed in the 1997 and 1998 'El Niño event' with yearly values of 4736 and 4790 mm. The lowest annual precipitation was observed in 1968,

FIGURE 6.2 Study area: sampling sites located in the upper (A) and middle (B) wetland areas. Fish sampling (C and D).

FIGURE 6.3 Some species of small littoral fish species of the Characidae family collected in Abras de Mantequilla wetland (*upper panel*). Species of commercial interest for local communities reported in the wetland (*low panel*). *Photos source Aguirre, W., 2014. The Freshwater Fishes of Western Ecuador. In: http://condor.depaul.edu/waguirre/fishwestec/intro.html.*

1975, and 2005, with values between 1066 and 1222 mm. The figure also highlights the years of the monitoring campaigns of the present study to relate them to historical rainfall variability.

The main inflow to the wetland is the Nuevo River that flows through the Estero Boquerón and contributes to 85% of total wetland inflow. During a strong rainy year like El Niño, inflow discharges from the Nuevo River to the wetland can reach maximum values up to 650 m^3/s, while during a dry year, maximum discharges are up to 260 m^3/s. The wetland also receives rainfall runoff from the Chojampe subbasin with a

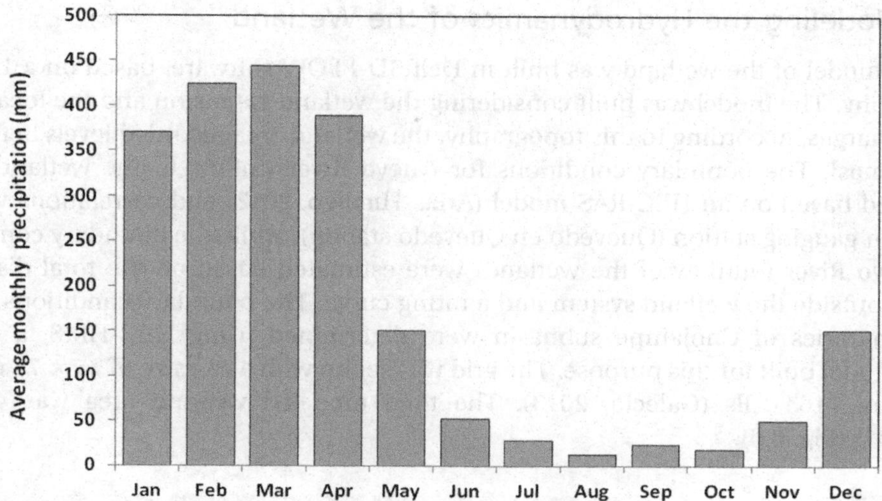

FIGURE 6.4 Average monthly precipitation in Quevedo-Vinces basin. Pichilingue station (1963–2012).

contribution of around 15% (Fig. 6.6). These contributions slightly fluctuate according to the type of year (dry or wet). During the dry season, the water level in the wetland decreases considerably, and water remains only in the deep central channels, reducing the inundated area to around 10% compared with the wet season.

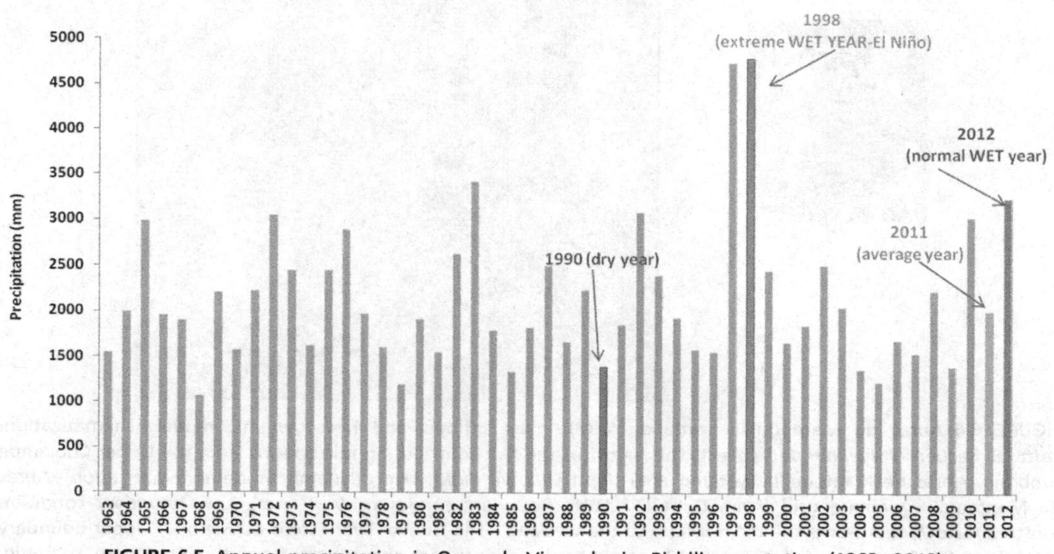

FIGURE 6.5 Annual precipitation in Quevedo-Vinces basin. Pichilingue station (1963–2012).

2.2 Modeling the Hydrodynamics of the Wetland

The 2D model of the wetland was built in Delft3D-FLOW software, based on a 1:10,000 topography. The model was built considering the wetland extension and the location of the discharges. According to this topography, the wetland area recorded levels between 6 and 34 masl. The boundary conditions for Nuevo River (inflow to the wetland) were estimated based on an HEC-RAS model (Arias-Hidalgo, 2012) and correlations with an upstream gauging station (Quevedo en Quevedo station), while the boundary conditions for Nuevo River (outflow of the wetland) were estimated based on the total discharge flowing outside the wetland system and a rating curve. The boundary conditions for the four tributaries of Chojampe subbasin were determined using HEC-HMS, a rainfall runoff model built for this purpose. The grid was set up with a cell size of 75 × 75 m, with a total of 7163 cells (Galecio, 2013). The total modeled wetland area was 4029 ha (40.3 km^2) (Fig. 6.6).

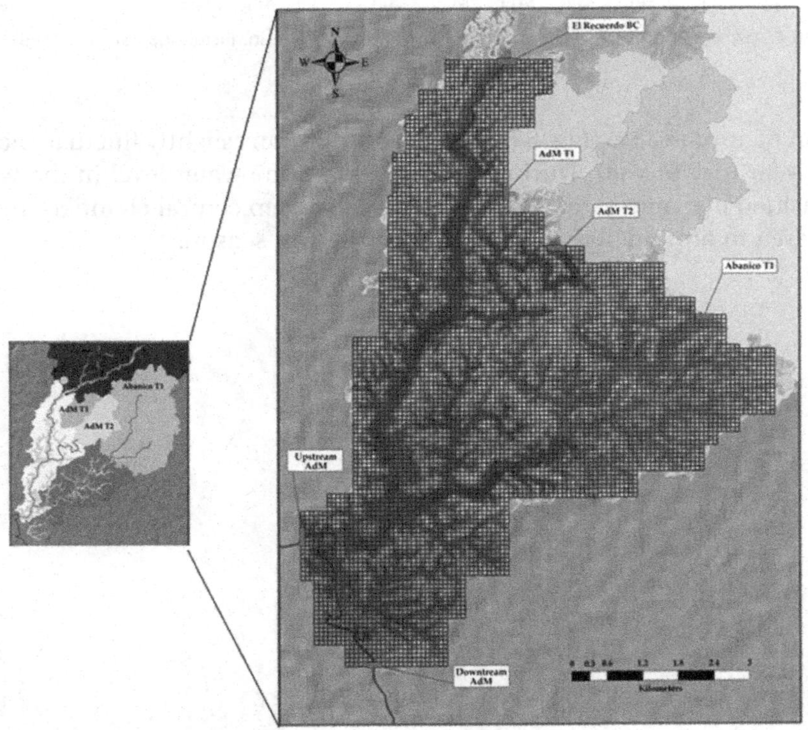

FIGURE 6.6 Abras de Mantequilla wetland (AdM)—main inflows and hydrodynamic model schematization. *Left*: "El Recuerdo" (*yellow dot*) collects the runoff of the five contributing microbasins from the Upper Chojampe subbasin. Abras de Mantequilla wetland area (*light yellow*). *Right*: Hydrodynamic model schematization—Abras de Mantequilla wetland grid (from Delft3D-FLOW). Boundary conditions (*red lines*). Low boundary condition (upstream AdM) represents the main inflow to the wetland "The Nuevo River-Estero Boquerón." Upper boundary conditions (El Recuerdo, AdMT1, AdMT2, and Abanico T1) collect the runoff from the Chojampe subbasin. *Source: Galecio, E., 2013. Hydrodynamic and Ecohydrological modeling in a tropical wetland: The Abras de Mantequilla wetland (Ecuador). UNESCO-IHE.*

Results from the hydrodynamic model show that the wetland is flooded up to 27 km^2 (Fig. 6.7). The natural variability of the wetland inundation area is depicted in Figs. 6.8 and 6.9. Monthly averages of inundated areas determine that historically the wetland experiences flooding from 5 to 23 km^2. A high variability between the different simulations is evident during the wet season. On the other hand, during the dry season, the inundation areas do not differ among the simulations, reaching all a value of 5 km^2 (Fig. 6.9). Nevertheless, the exception is the maximum historical condition since this time series includes the complete set of extreme wet conditions for a long historical period (1962–2010). As an example to illustrate both spatial and temporal variation in inundation patterns, water depth maps from 2012 are presented in Fig. 6.10.

2.3 The Habitat Suitability Index

The aim to develop a habitat index (HI) is to indicate how suitable an area is for a determined species or group of species. Nevertheless, it has a number of assumptions, for example, it is not clear if an index will certainly indicate the presence or absence of either these species or the quantity of the species. On the other hand, to be able to determine a species habitat it is important to know in which period a species distributes and if this period is logical. Thus by exploring the space–time variability of an index the presence of these species could be estimated. The hydrological behavior of the wetland explained in the previous section provides the base for our habitat suitability approach, given that the hydrology shapes our habitat. The following section of the methodology describes the multiple tests performed to determine the level of relationship between water depth, velocity, and habitat for the fish community in this tropical wetland:

1. Calculation of the habitat index (HI) was based on a general rule based on literature of the natural behavior of fish in the study area. Some key aspects explored include seasonal behavior, spawning, food availability, and inundation area fluctuations. Field information and expert knowledge were used to validate this information. In situ measurements of water depth and velocity were compared with literature values to understand the distribution and habitat preferences of the overall fish community. As a result, response curves (knowledge rules) for these two variables were derived (Fig. 6.11).
2. A dynamic HABITAT modeling tool was built with the MATLAB toolbox (Fig. 6.12).
3. Habitat suitability was evaluated by relating in situ measurements of water depth and flow velocity from sampling years 2011 and 2012 with the results of the 2D hydrodynamic model (Delft3D-FLOW). Furthermore, extreme conditions (dry and wet years) and minimum and maximum historical conditions were also modeled to account for natural variability.
4. Output maps of water depth and velocity from the 2D hydrodynamic model (Delft3D-FLOW) were used as an input for the dynamic HABITAT model.

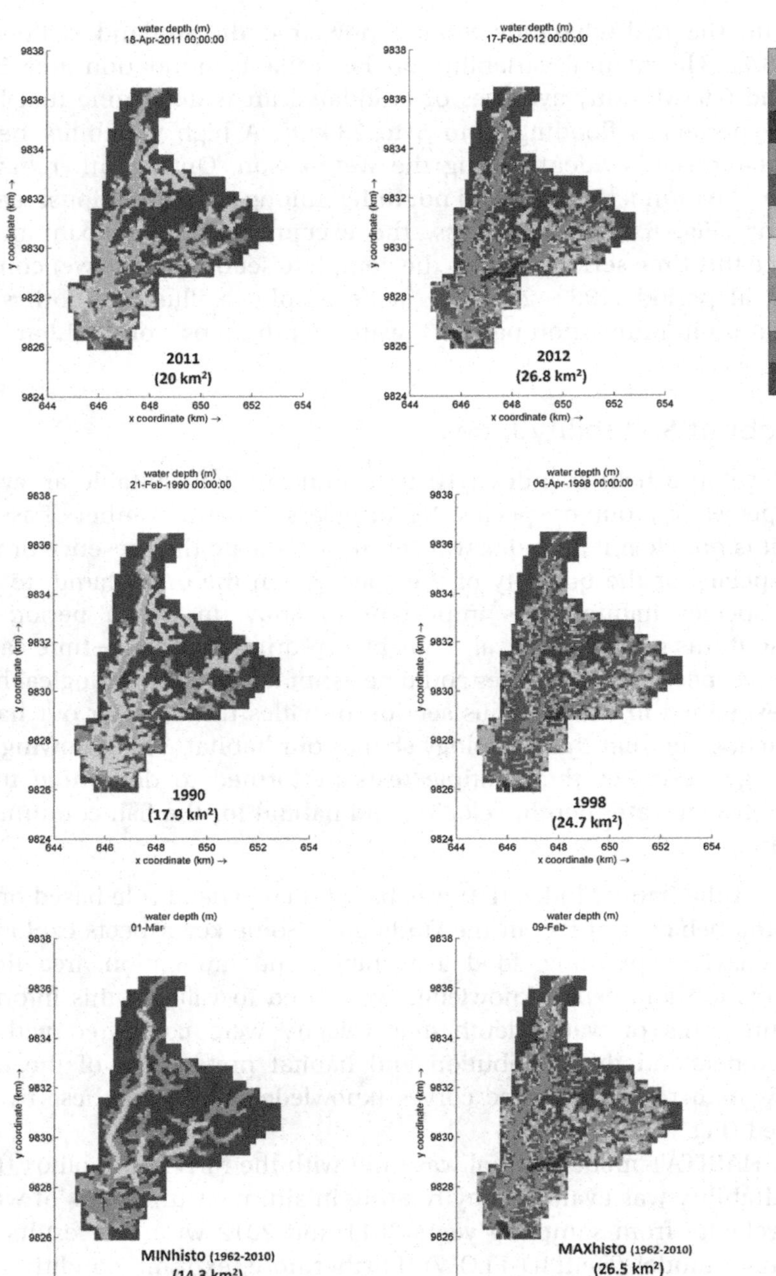

FIGURE 6.7 Maximum wetland inundated areas (km^2) from Delft3D-FLOW simulations for: 2011 and 2012 (sampling years); 1990 (dry year); 1998 (wet year); and minimum and maximum historical conditions (period 1962–2010). Scale bar indicates the water depth range (m).

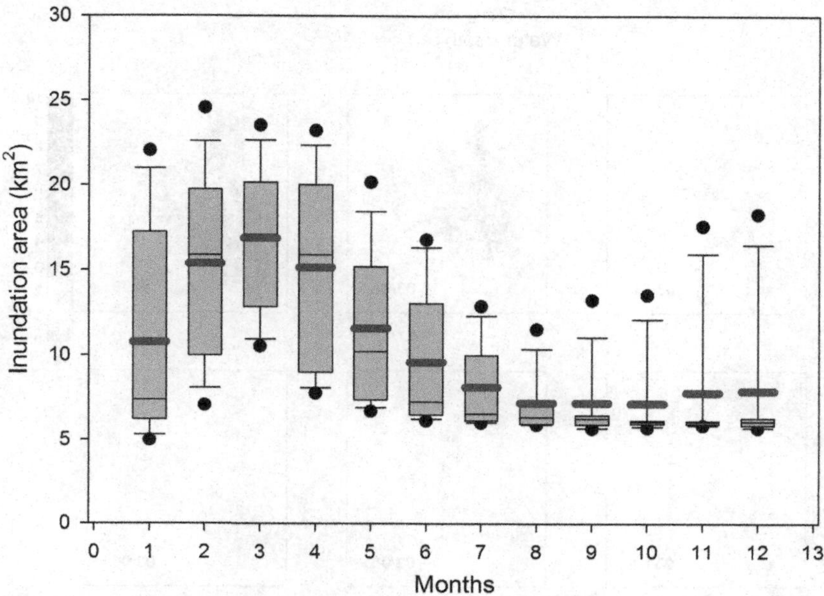

FIGURE 6.8 Boxplot for the wetland inundation area (km^2) calculated from Delft3D-FLOW simulations results: 1990 (dry year); 1998 (wet year); 2011 and 2012 (sampling years); and minimum and maximum historical conditions (period 1962–2010). Each month is built with the daily values of each simulation. *Blue line* (mean), *upper and lower dots* (5th and 95th percentiles).

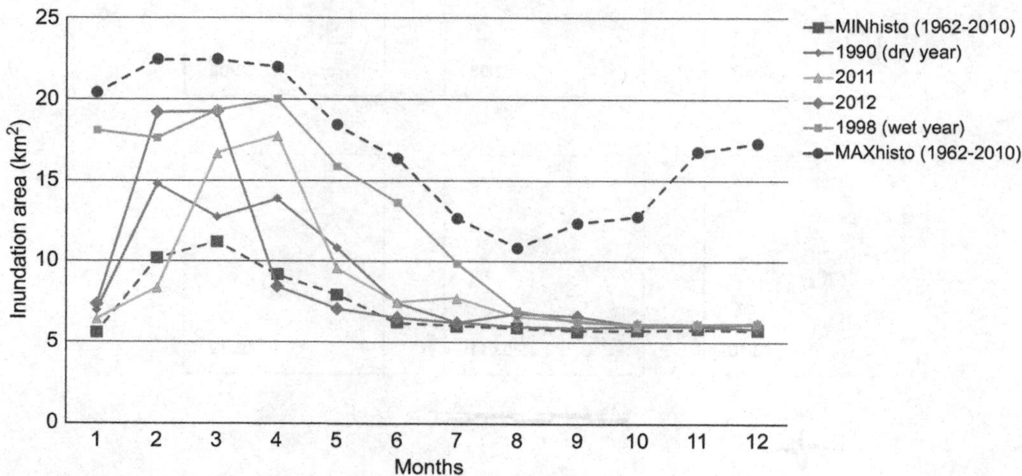

FIGURE 6.9 Monthly average of wetland inundation area (km^2). Built from Delft3D-FLOW simulation results.

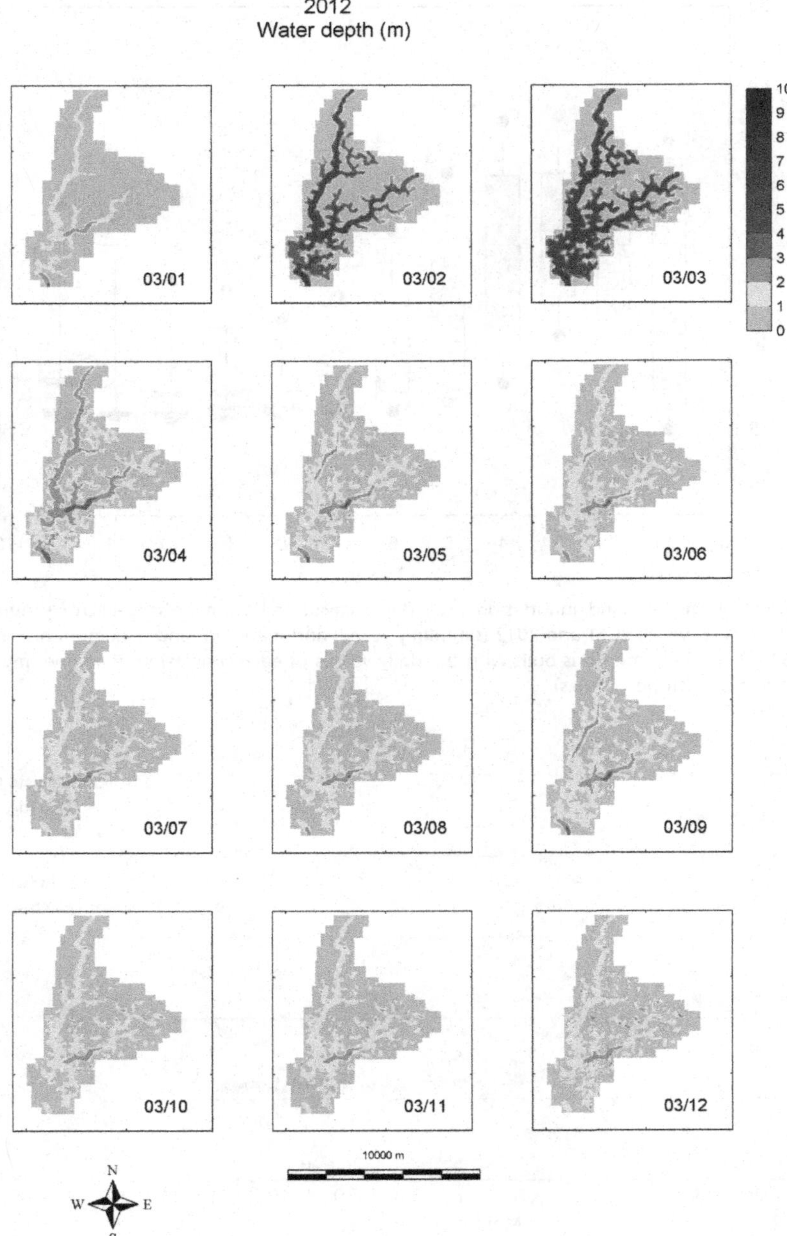

FIGURE 6.10 Water depth (m) maps (January–December 2012) from DELFT- FLOW. Output maps extracted the same day every month. Scale bar indicates the water depth range (m). Months display: top left (January), low right (December).

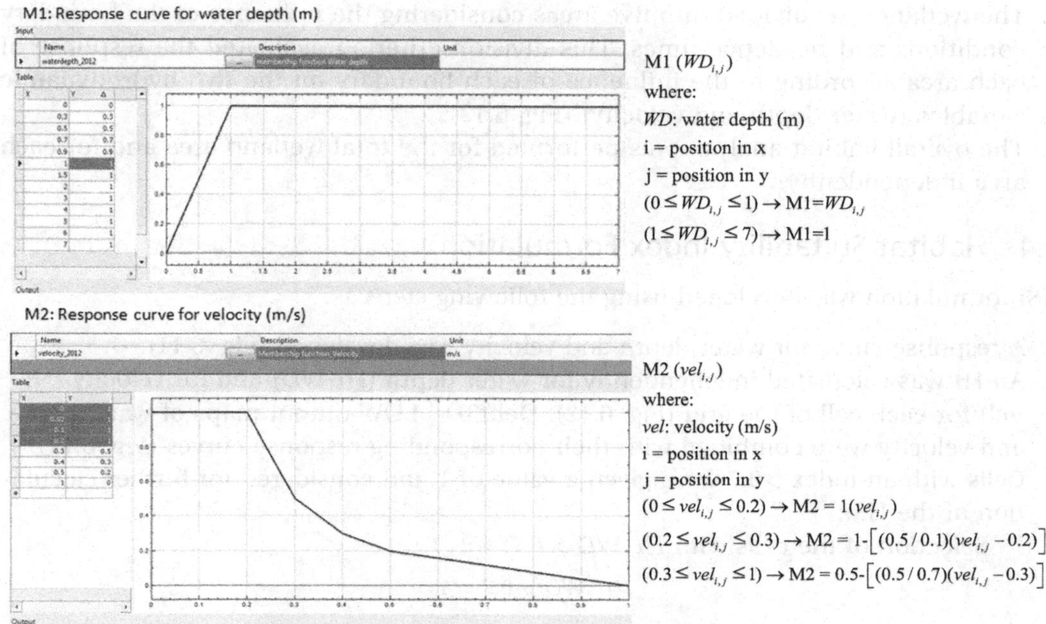

M1: Response curve for water depth (m)
M1 ($WD_{i,j}$)
where:
WD: water depth (m)
i = position in x
j = position in y
$(0 \leq WD_{i,j} \leq 1) \rightarrow M1 = WD_{i,j}$
$(1 \leq WD_{i,j} \leq 7) \rightarrow M1 = 1$
M2: Response curve for velocity (m/s)
M2 ($vel_{i,j}$)
where:
vel: velocity (m/s)
i = position in x
j = position in y
$(0 \leq vel_{i,j} \leq 0.2) \rightarrow M2 = 1(vel_{i,j})$
$(0.2 \leq vel_{i,j} \leq 0.3) \rightarrow M2 = 1\text{-}\left[(0.5/0.1)(vel_{i,j} - 0.2)\right]$
$(0.3 \leq vel_{i,j} \leq 1) \rightarrow M2 = 0.5\text{-}\left[(0.5/0.7)(vel_{i,j} - 0.3)\right]$

FIGURE 6.11 Response curves for water depth (M1) and velocity (M2). The *x* axis presents the variable values: water depth (m), velocity (m/s); the *y* axis presents the habitat index score.

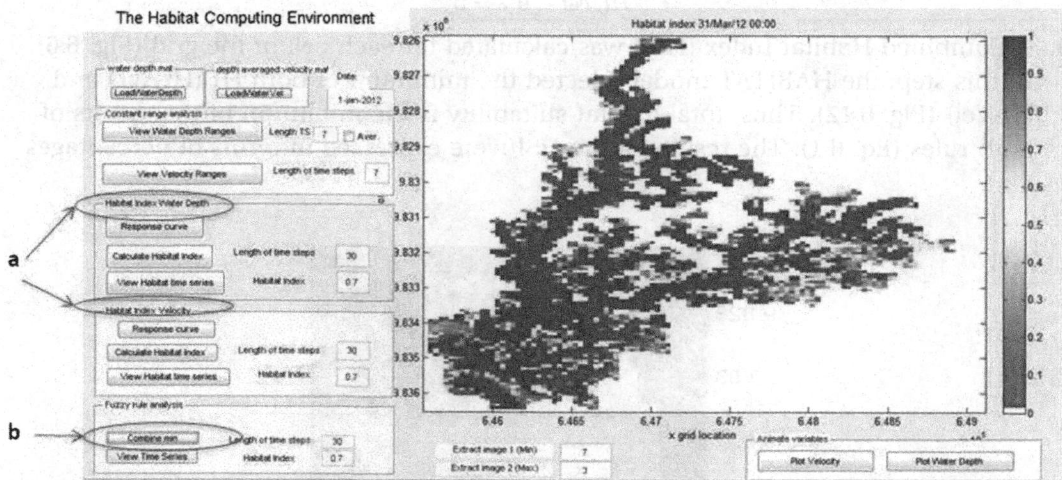

FIGURE 6.12 The dynamic habitat computing tool. (a) Constant range analysis, (b) variable habitat index (VHI), (C) combined habitat index (Comb-HI). Color bar indicates the habitat index scale (0: not suitable; −1: most suitable).

5. The wetland was divided into five areas considering the influence of the boundary conditions and residence times. This division criterion evaluated the response of each area according to the influence of each boundary on the two hydrodynamic variables (water depth and velocity) (Fig. 6.13).
6. The overall habitat analysis was performed for the total wetland area and for each area independently.

2.4 Habitat Suitability Index Formulation

HSI formulation was developed using the following steps:

1. A response curve for water depth and velocity was developed (Fig. 6.11).
2. An HI was calculated independently for water depth (HI-WD) and for velocity (HI-Vel) for each cell of the grid (Fig. 6.12). Delft3D-FLOW output maps of water depth and velocity were combined with their corresponding response curves (Fig. 6.11). Cells with an index >0.7 were given a value of 1 and considered for further calculation of the HSI:

   Selection of the cells with $HI\_WD > 0.7$:

$$HI\_WD > 0.7 = 1$$

$$HI\_WD < 0.7 = 0.$$

   Selection of the cells with $HI\_vel > 0.7$:

$$HI\_vel > 0.7 = 1$$

$$HI\_vel < 0.7 = 0.$$

3. A Combined Habitat Index (HSI) was calculated for each cell of the grid (Fig. 6.6). In this step, the HABITAT model selected the minimum of both HI (HI-WD and HI-Vel) (Fig. 6.12). Thus, total habitat suitability is the minimum of the results of both rules (Eq. 6.1). The results of the HSI were expressed in terms of percentages

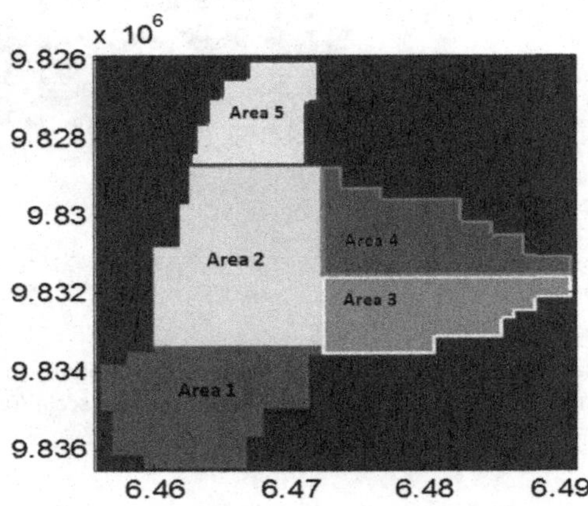

FIGURE 6.13 Wetland areas delimitation.

of suitable areas (PSA) with HSI >0.7. PSA was calculated for each time step. (Eq. 6.2):

$$HSI = \mathrm{Min}(HI\_WD, HI\_vel) \tag{6.1}$$

$$PSA = \frac{\sum_{k=1}^{n} HSI \geq 0.7}{N} \times 100 \tag{6.2}$$

Where:

HSI = Habitat Suitability Index

n = each cell

N = total number of cells

Furthermore, in order to have a large scale and overall HSI for each of the five areas, a second approach was applied. In this second approach, the cell values of water depth and velocity of each area were averaged before the calculation of *HI_WD* and *HI_vel*. Subsequently, the HSI was calculated in the same way as the previous approach by selecting the minimum of both. Results of this second approach were expressed in HSI (with a scale from 0 to 1), both temporally and spatially. All the calculations for approaches 1 and 2 were performed for the total wetland area, and also for each of the five areas independently.

3. Results

Two different approaches were evaluated to explain the habitat conditions of this tropical wetland. Results of the first approach are expressed in terms of PSA with an HSI above 0.7 (Sections 3.1–3.3). A second approach evaluated the wetland in terms of HSI scores (Sections 3.4–3.6).

3.1 Natural Variability of Suitable Areas

Our analysis started by evaluating the temporal distribution of the PSA with an HSI above 0.7 (first approach). Different hydrological years were simulated to understand the natural variability. Results described a high variation in terms of suitable areas depending on the hydrological conditions simulated. During a dry year, the percentage of suitable areas was up to 40% of the total wetland area, increasing to around 70% during wet years and historical maximum condition. Sampling years 2011 and 2012 were between both extreme conditions, with 2012 presenting higher percentages of suitable areas (up to 60%) compared to 2011 (up to 50%). Minimum (MINhisto) and maximum (MAXhisto) temporal distributions provided with the limits to understand the historical thresholds that the wetland has experienced during the period 1962-2010. The simulation of the minimum time series shows that historically the wetland had always provided at least a 25% of suitable area even in this extreme condition. For all conditions, higher percentages of suitable areas occurred during the wet season (January–May) (Fig. 6.14).

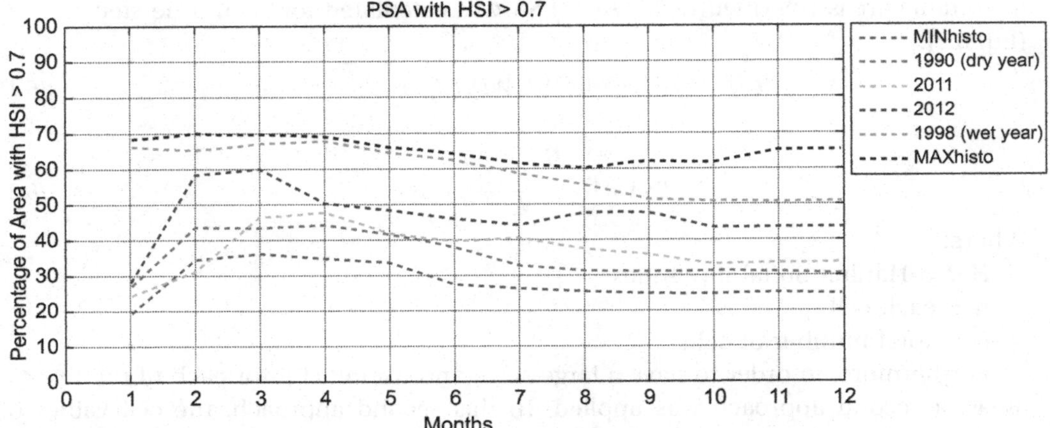

FIGURE 6.14 Temporal distribution of the percentage of suitable wetland area (PSA) with Habitat Suitability Index (HSI) >0.7 for sampling years (2011 and 2012), compared with extreme dry and wet years (1990 and1998), and minimum and maximum historical conditions (period 1962–2010). Months: January (1) to December (12).

3.2 Contribution of Each Wetland Area to the Total Wetland Suitable Area

Fig. 6.15 illustrates the contribution of each of the five areas to the total wetland area with an HSI > 0.7. From the results, it can be seen that areas 1 and 2 have a higher contribution, while the rest of the areas contribute less. The proportion of this contribution is maintained throughout the years analyzed. The timing at which the maximum of suitable areas occurred during the wet season differed between the years. Thus in 2011 it occurred during March and April, and in 2012 during February and March.

3.3 Independent Analysis of PSA per Area

Each of the five wetland areas was also analyzed independently in terms of PSA. For this analysis each area was compared to its own total area. Fig. 6.16 illustrates the temporal behavior of each area for sampling years 2011 and 2012. From the analysis, it is shown that areas 1 and 2 are the ones with a higher percentage of suitable areas HSI > 0.7, with percentages up to 70% and 50% in 2011, and 80% and 65% in 2012, respectively. Wetland area 3 showed an intermediate behavior during both years, with percentages around 45% in 2011 and 60% in 2012. Lower percentages were observed for areas 4 and 5. Area 4 showed values around 30% in 2011 and up to 40% in 2012, showing a clear separation from area 5 in 2012, while in 2011 both areas followed a similar pattern.

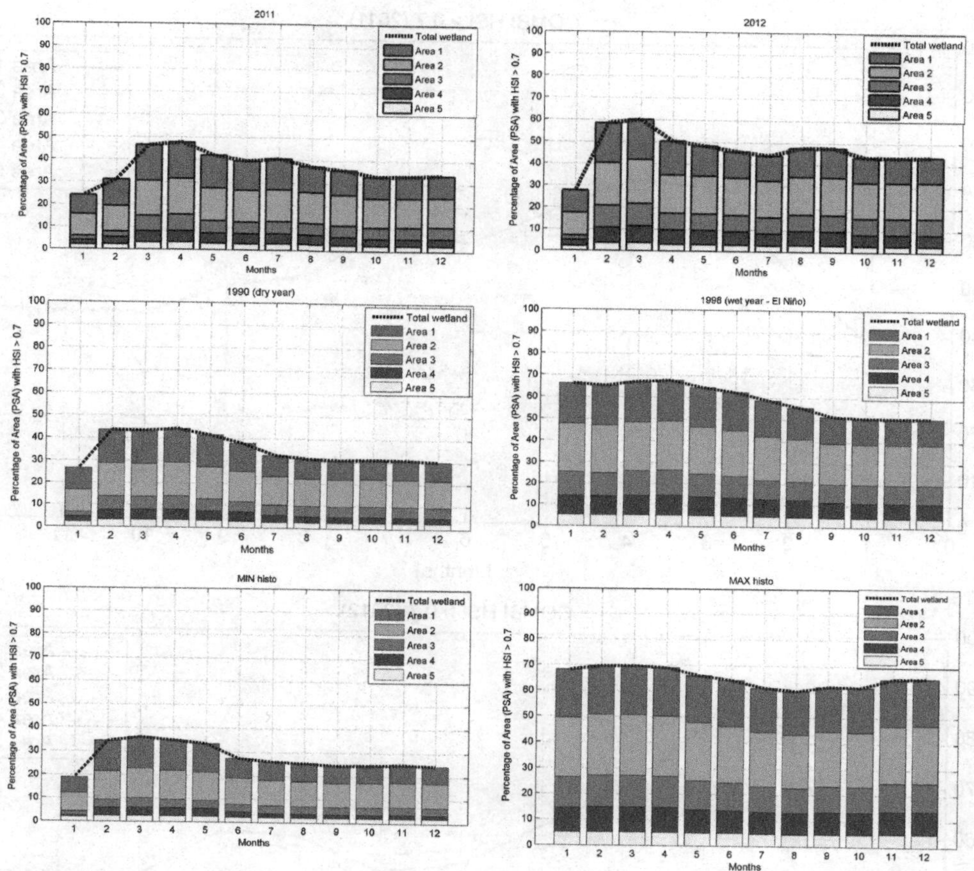

FIGURE 6.15 Contribution of each wetland area to the total wetland area with Habitat Suitability Index (HSI) > 0.7. For sampling years (2011 and 2012) (*upper panel*), dry and wet (1990 and1998) (*middle panel*), and minimum and maximum historical conditions (1962–2010) (*lower panel*). The sum of the five areas is equal to the total wetland area (*dashed black line*). Months: January (1) to December (12).

3.4 Natural Variability of the HSI

The second approach of this chapter analyzes the wetland in terms of HSI scores. As well as in the first approach, different hydrological years were simulated to understand the natural variability of the index. The dry year maintained an HSI score of 1 from February to May, decreasing to 0.5 from July on. On the other hand, a wet year (El Niño) maintained an HSI of 1 for a longer period, decreasing slightly to 0.8 from September on. Sampling years 2011 and 2012 reached the maximum score in the months of

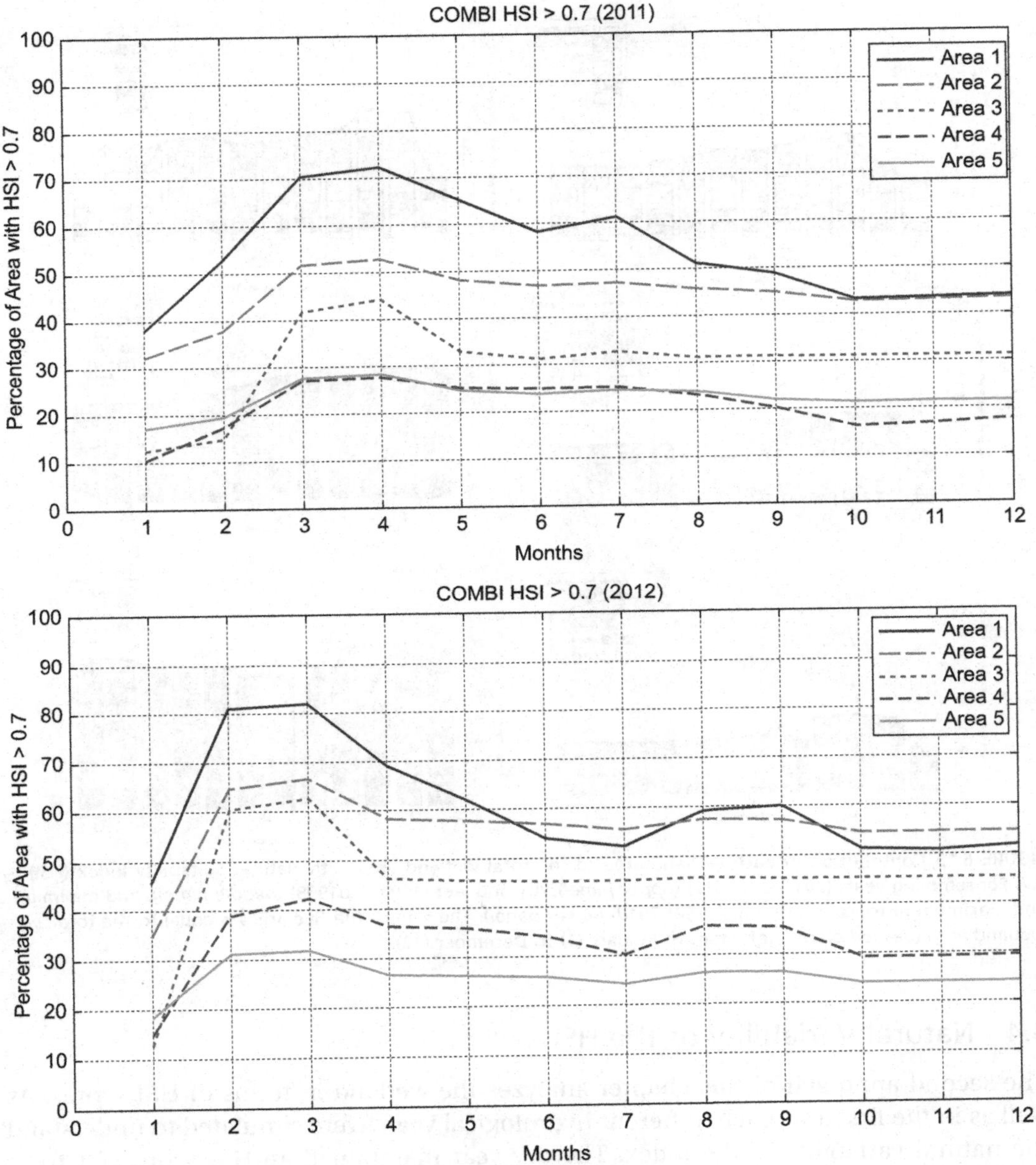

FIGURE 6.16 Percentage of suitable area (PSA) with a Habitat Suitability Index (HSI) >0.7 for each wetland area. Sampling years 2011 and 2012. Months: January (1) to December (12).

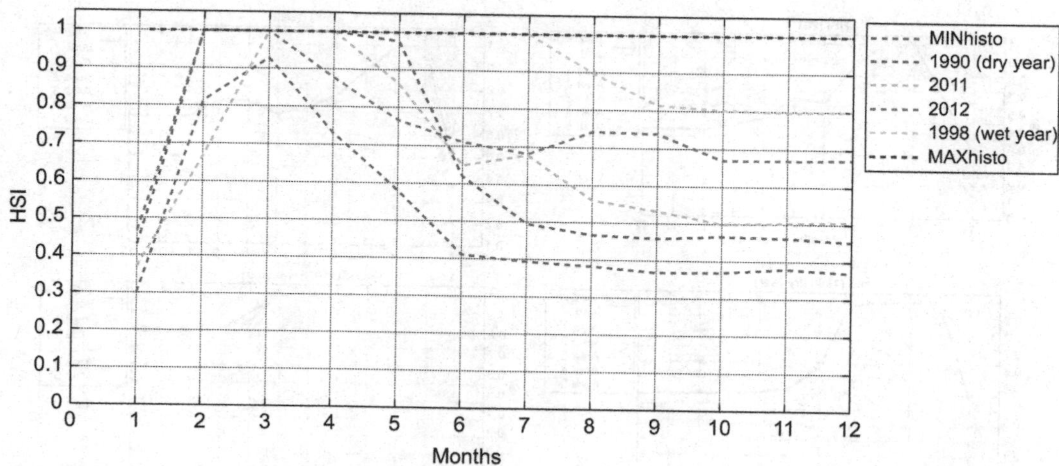

FIGURE 6.17 Temporal distribution of the Habitat Suitability Index (HSI) for sampling years (2011 and 2012), dry year (1990), wet year (1998), and minimum and maximum historical conditions (period 1962–2010). Months: January (1) to December (12).

March–April and February–March, respectively. The year 2011 followed a similar trend of a dry year, with values around 0.5 during the dry season. Extreme scenarios indicated that during maximum conditions the wetland can maintained an HSI of 1 during the whole year. The time series of the minimum historical condition illustrated that in the most unfavorable conditions, the HSI score in the wetland was around 0.4 (Fig. 6.17).

3.5 Independent Analysis of the HSI per Area

The five wetland areas were also analyzed independently in terms of HSI scores. Fig. 6.18 illustrates the temporal behavior of the HSI in each area for the different hydrological conditions. From this analysis, areas 1 and 2 were the ones with higher HSI scores; area 3 exhibited intermediate scores; while areas 4 and 5 had the lowest scores, for all the simulated conditions. During a dry year, a clear separation between the areas was observed along the whole simulation period, while during a wet year this separation was only evident during the dry season period. The maximum historical condition displayed a constant highest score of 1 during the whole simulation, with the exception of areas 4 and 5 that slightly decreased to 0.9 during August. It was interesting to see that higher scores for areas 1 and 2 were also reached during the minimum historical simulation. Overall and from a temporal perspective, all wetland areas reached higher scores of HSI during the wet season period.

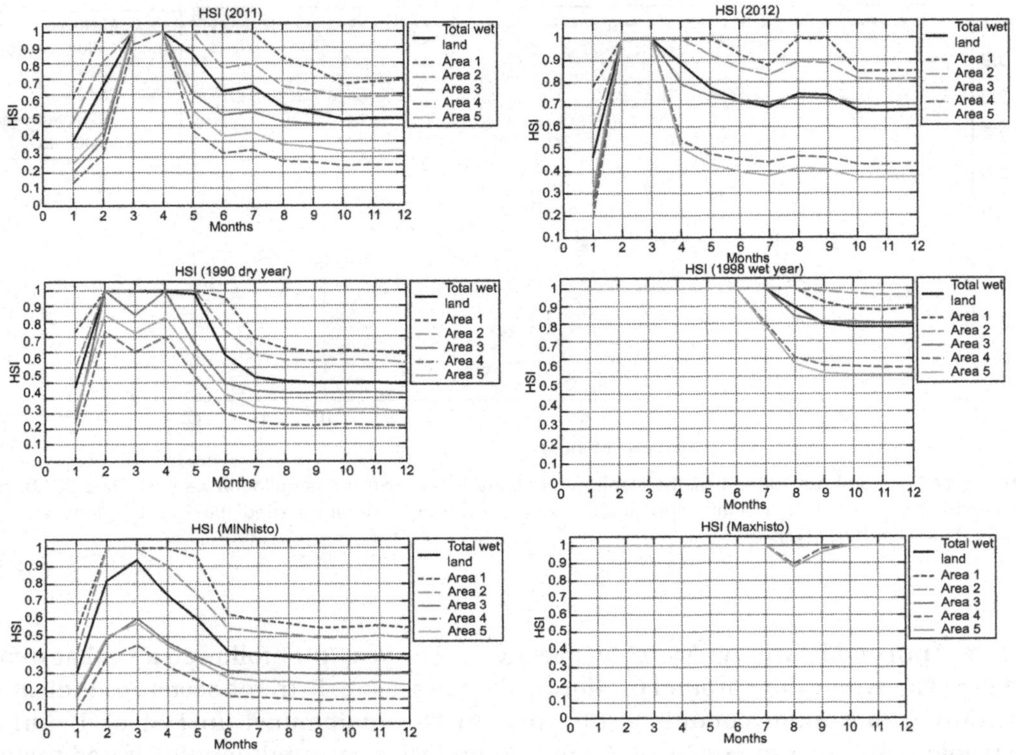

FIGURE 6.18 Temporal distribution of Habitat Suitability Index (HSI) for each wetland area (*colored lines*) and total wetland area (*dashed black line*) for sampling years (2011 and 2012), dry year (1990), wet year (1998), and minimum and maximum historical conditions (period 1962–2010). Months: January (1) to December (12).

3.6 Spatial and Temporal Variation of the HSI

Fig. 6.19 displays the spatial and temporal variation of the habitat suitability index (HSI) for the different hydrological conditions. During the first 6 months of a wet year (El Niño year 1998), all areas reached an HSI of 1, and were >0.6 even during the dry season. The maximum historical simulation showed a constant HSI above 0.9 during the whole year for all the areas. Overall, spatial HSI results for simulations were 1990, 2011, 2012, and minimum historical described areas 1 and 2 as the ones with higher HSI scores (even during the dry season [>0.6]), while areas 4 and 5 were the ones with the lowest HSI scores (HSI values <0.5 during the dry season). Area 3 presented intermediate HSI values. Thus results suggest that wetland areas 1 and 2 are the ones that provide better conditions for fish. Since wetland areas 4 and 5 showed lower HSI scores, these areas may require special attention in terms of management. Temporal behavior of the HSI (for all simulations) defined the wet season period (February–April) as the key period providing suitable habitat conditions for the entire fish community of Abras de Mantequilla wetland.

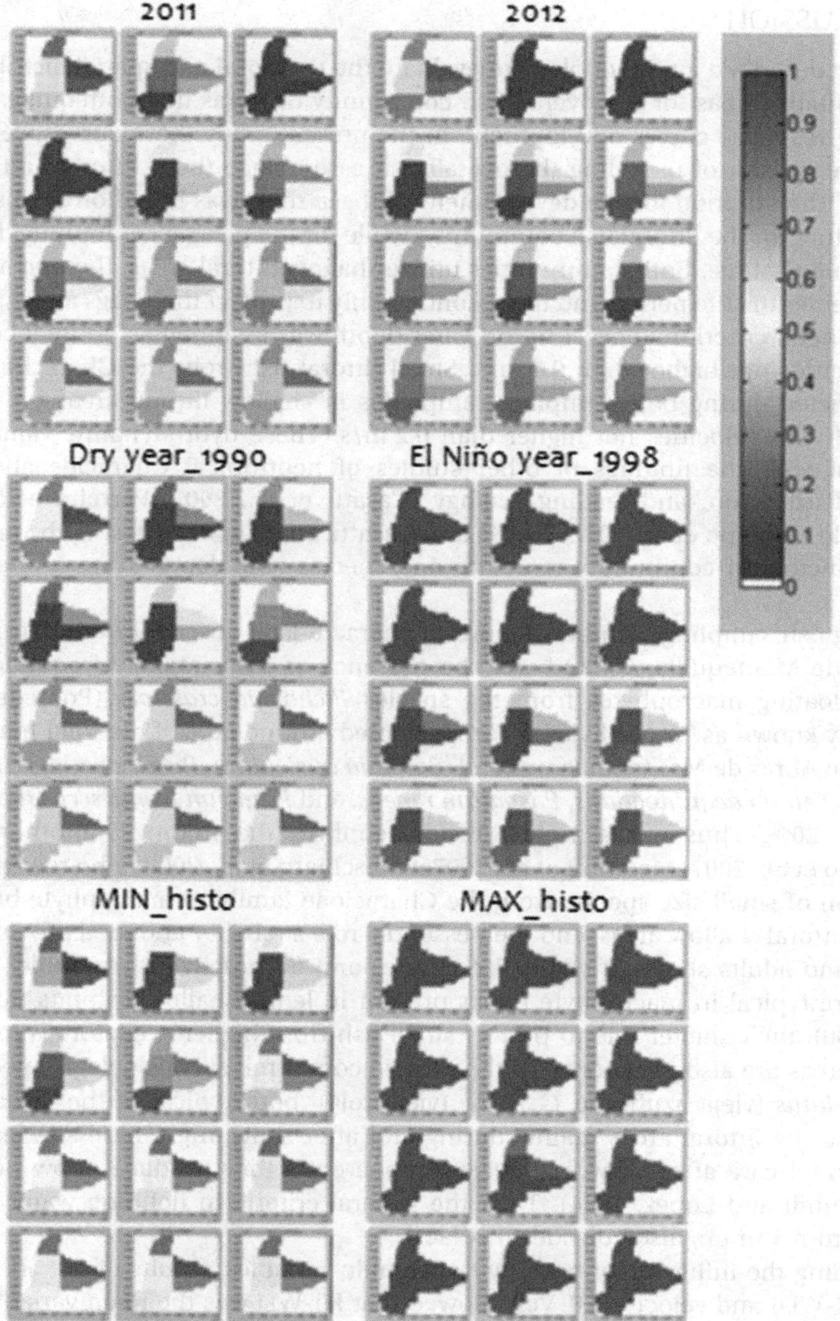

FIGURE 6.19 Spatial and temporal distribution of Habitat Suitability Index (HSI), for sampling years (2011 and 2012), dry year (1990), wet year (1998), and minimum and maximum historical conditions (period 1962–2010). Months display follows the same sequence as Figure 6.10: top left (January), low right (December). *Color bar* indicates the habitat index scale from 0 (not suitable in *red*) to 1 (most suitable in *dark green*).

4. Discussion

This study described a methodology to evaluate the temporal and spatial distribution of habitat suitable areas for the overall fish community of Abras de Mantequilla wetland. Regarding response curves development, in the present study these curves were developed with the aim of including the overall fish assemblage (both littoral and pelagic/limnetic). The criterion for the development of these rules was based on field sampling and literature for the littoral fish community, while for the pelagic community, literature was the main source. Both communities utilize shallow littoral areas, the first as habitat during their entire life period and the second mainly to protect their eggs after spawning. Thus a general criterion above 1 m for water depth was assumed as optimal combined with velocities not higher than 0.2 m/s. Small littoral fish from the Characidae family were collected during both sampling campaigns in shallow littoral areas up to 1.5 m, combined with velocities not higher than 0.2 m/s. These hydrodynamic values are in agreement with the findings of other studies of neotropical Characids about their habitat, distribution, and feeding ecology (Casatti et al., 2003; Ferreira et al., 2012; Maldonado-Ocampo et al., 2012; Teresa and Casatti, 2013) and suggest both variables as good predictors of community structure and species abundance (Teresa and Casatti, 2013).

During fish sampling, another important characteristic observed in the littoral areas of Abras de Mantequilla wetland was the presence of associations of aquatic macrophytes. Floating macrophytes from the species *Eichornia crassipes* (Pontederiaceae), commonly known as "water hyacinth" represented around 80% of the total macrophyte biomass in Abras de Mantequilla wetland. *Salvinia auriculata, Pistia stratiotes, Ludwigia peploides, Lemna aequinoctialis, Paspalum repens,* and *Panicum frondescens* represented the other 20%. Thus our sampling results confirm the findings of other authors (Agostinho et al., 2007; Meerhoff et al., 2007a; Meschiatti et al., 2000) who recognized the association of small size species from the Characidae family to macrophyte banks that colonize littoral shallow areas, and their essential role as shelter and food provider. Since juvenile and adults stages of small size species and eventually also juveniles of larger species are typical in macrophyte banks present in lentic shallow habitats (Meschiatti et al., 2000), their shelter role to protect small fish from higher predators is important. Shallow areas are also important for the pelagic community. Thus pelagic species such as *A. rivulatus* (vieja azul) and *C. festae* (vieja roja), both typical of the wetland area, utilize also the littoral areas mainly during and after spawning. These species present high parental care after spawning because apparently they produce a low number of eggs (Barnhill and Lopez, 1974). Thus the general criteria in defining water depth as optimal from 1 m on, also consider this fact.

Regarding the influence of both hydrodynamic variables, results of the HI for water depth (HI-WD) and velocity (HI-Vel) showed that HI-WD was the main variable driving the HSI results because velocities were quite homogeneous in the entire wetland area.

Our findings revealed a high natural variability of the of the percentage of suitable areas (PSA) according to the different hydrological conditions simulated. Thus, from the historical perspective the results showed that the wetland can provide a range between 25% and 70% of suitable areas (given the response curves implemented for this study). These limits can be used as minimum and maximum thresholds for management purposes. Regarding temporal behavior, results describe the wet season period (January–May) as the one with a higher PSA for all simulations. Nevertheless, during extreme wet conditions, a higher PSA was also observed during what is considered normally the months of the dry season period.

Spatial analysis described areas 1 and 2 as those that provided more suitable habitat conditions and contributed higher percentages to total wetland habitat suitability. These areas are the ones that fulfill best the conditions described in the response curves. Local physical characteristics of these areas, as topography and proximity of the main inflow (Nuevo River), appeared to be the main drivers of our results. This is in agreement with the higher catch per effort for fishing activities reported in San Juan de Abajo (Florencio, 1993). This location belongs to area 1 of our study (low wetland area). On the other hand, areas 4 and 5 related to the Upper Chojampe inflows were the ones with a lower percentage of suitable areas. In these areas, the main source of water is related to runoff, and not to river inflow. Thus these areas will require specific management measures in the future to maintain their inflow contribution.

When the wetland was evaluated in terms of HSI scores, a similar pattern was observed for both spatial and temporal results. Thus higher HSI scores were obtained for areas 1 and 2 despite the hydrological condition simulated, and in general the months corresponding to the wet season period were the ones exhibiting higher scores for all simulations. From the historical perspective, and considering the whole wetland area, HSI scores were not lower than 0.4, even in the most unfavorable conditions (minimum historical).

This temporal availability of suitable areas definitely plays an important ecological role in the basin, since the majority of the fish of the Vinces River and associated floodplains present one reproductive cycle per year. At the end of the dry season, several fish species have a mature state ready for spawning. These species usually have high fecundity (high number of eggs to assure an adequate repopulation). However, there are also species such as *Aequidens rivulatus* (vieja azul) and *Cichlasoma festae* (vieja roja) that spawn during the transition periods between the wet and dry seasons, and others such as *Brycon dentex* that have been reported in mature stages also during the dry season (Barnhill and Lopez, 1974). Thus both seasons are important but for different species, therefore the importance of maintaining the natural timing of the inflows. A study on the biological aspects of the fish community in the basin revealed that 70% of the specimens sampled during the months of January to March reported an advanced stage of sexual maturity (stages III–V) (Revelo, 2010), and the rest of the specimens were already in stages of postspawning (I and II). When the analysis was more specific per species, the timing of mature stages differed slightly between the months of the wet season. For instance, *B. dentex* (dama) reported a higher number of specimens with an advanced mature stage in January (III, IV, and V), while in February and March immature stages were more frequent, indicating that spawning probably occurred between January and February. *Ichthyoelephas humeralis* (bocachico) reported specimens in advanced mature stages (III and IV) during January and February. Other species of less commercial interest such as *Hoplias microlepis* (guanchiche) reported mature stages during the first 3 months of the year (January–March), and immatures during April. *A. rivulatus* (vieja azul) reported specimens in advanced maturity stage (III, IV, and V) during January and February, and immature stages in April and May. Also a smaller percentage of immatures was reported during October and November, possibly explaining that this species has two reproductive cycles per year. *Curimatorbis boulengueri* (dica) reported advanced mature stages during February, and immature stages during March and April (Revelo, 2010). The last one is confirmed by the sampling performed during the present research where small immature specimens of *C. boulengueri* were collected during March 2012. All these findings provided evidence that the wet season and associated high flows represent an important period for the development and increase of the fish population in the study area, which is consistent with findings of other tropical systems that acknowledge the importance of high flows and floods in supporting the gonadal maturation of fish (McClain et al., 2014).

Fishing activities in the wetland occurred during 10 months of the year, but start usually at the end of the wet season (Florencio, 1993). Local farmers from 'El Recuerdo' village have also reported the catching of bigger size fish during the dry season (T. Estrella, pers. comm., 2016). They mentioned that they wait until the sizes are big enough to catch them to allow the fish population to grow. In this regard, there is also a regulation in Los Rios Province that establishes a ban ("veda") for fishing activities from January 10 to March 10 (Revelo, 2010).

A parallel study on biotic communities structure (Alvarez-Mieles et al., 2018) determined that although the different areas of the wetland shared similar fish species from the Characidae family, there were species that seem to typify middle areas where higher residence times take place. Other species typify the lower area, which is more influenced by the river inflow, while other species characterize the river inflow. These findings at species level can provide the basis for future research on the construction of new rules and habitat assessment at a lower taxonomic level for the fish community in this tropical wetland.

5. Conclusions

The present study is the first attempt at providing an assessment of the temporal and spatial variation of suitable habitat areas for the fish community in Abras de Mantequilla wetland. Our results evaluated how hydrodynamic variables can facilitate the definition of suitable habitat areas in this wetland, in terms of both PSA and HSI scores. In this study, wetland areas with HSI >0.7 were described as optimal habitat for this fish community. However, areas with an HSI <0.7 cannot be considered necessarily as uninhabitable.

One of the limitations of our approach is that the rules were developed for the whole fish community, rather than for specific species. For this, more extensive sampling in the area is required to measure the habitat preference of different species. However, our methodology can provide an initial base for future habitat assessments of specific fish communities in the area.

The combination of hydrodynamic variables proved to be useful for an initial habitat assessment of the fish communities in this wetland. However, we acknowledged that other physical, chemical, and biotic variables play an important role in defining the habitat preferences and therefore should be gradually included for an integrated ecological habitat assessment. In this regard, the habitat tool developed for this study is quite flexible for adding more variables and their corresponding rules.

The high flow phase of the wet season was recognized as the period with a higher percentage of suitable areas and HSI scores for all the simulated conditions. Spatial zonation defined the areas close to the main inflow as the ones providing better habitat conditions, and areas related to the Chojampe subbasin as the ones that will require special attention in terms of management measures.

Based on the results of this study, it is recommended to maintain the timing and magnitude of the natural flows especially during periods with a higher percentage of suitable areas (high flows of the wet season), since this period is crucial to foment the spawning and development of the fish community in this wetland.

Acknowledgments

This study is part of the Ph.D. research of the first author (funded by NUFFIC-The Netherlands Fellowship Programme). The sampling campaign was sponsored by WETwin (EU Project). The National Institute of Fisheries (INP) in Guayaquil-Ecuador provided support with the logistics and secondary information of the area. Thanks to INP staff for their cooperation in the plankton and water quality sampling (results not included in this study but they are also part of the Ph.D. thesis). Fish sampling was possible due to the valuable collaboration of Antonio Torres from Facultad de Ciencias Naturales-Universidad de Guayaquil. Taxonomical identification of the collected fish specimens was developed by Antonio Torres in collaboration with students at the Laboratory of Aquaculture (Universidad de Guayaquil). Thanks to Windsor Aguirre, Assistant Professor at DePaul University of Chicago, for his key opinions about the status of Ecuadorian freshwater fishes and to Enrique Galecio for developing his master's thesis in Abras de Mantequilla, and his valuable work setting up the hydrodynamic model. Special thanks to local inhabitants Telmo España, Jimmy Sanchez, and Simon Coello for boat transportation and for sharing their valuable local knowledge.

References

Aarts, B.G.W., Van Den Brink, F.W.B., Nienhuis, P.H., 2004. Habitat loss as the main cause of the slow recovery of fish faunas of regulated large rivers in Europe: the transversal floodplain gradient. River Research and Applications 20 (1), 3–23. https://doi.org/10.1002/rra.720.

Agostinho, A., Thomaz, S., Gomes, L., Baltar, S.S.M.A., 2007. Influence of the macrophyte Eichhornia azurea on fish assemblage of the Upper Paraná River floodplain (Brazil). Aquatic Ecology 41 (4), 611–619. https://doi.org/10.1007/s10452-007-9122-2.

Aguirre, W., 2014. In: The Freshwater Fishes of Western Ecuador. http://condor.depaul.edu/waguirre/fishwestec/intro.html.

Aguirre, W.E., Shervette, V.R., Navarrete, R., Calle, P., Agorastos, S., 2013. Morphological and genetic divergence of *Hoplias microlepis* (characiformes: erythrinidae) in rivers and artificial impoundments of western Ecuador. Copeia 2013 (2), 312–323.

Alvarez-Mieles, G., Irvine, K., Griensven, A.V., Arias-Hidalgo, M., Torres, A., Mynett, A.E., 2013. Relationships between aquatic biotic communities and water quality in a tropical river–wetland system (Ecuador). Environmental Science & Policy 34 (0), 115–127. https://doi.org/10.1016/j.envsci.2013.01.011.

Alvarez-Mieles, G., Irvine, K., Mynett, A.E., 2018. Biotic Communities Structure under Two Different Inundation Conditions (in preparation).

Arias-Hidalgo, M., 2012. A decision framework for integrated wetland-river basin management. In: A Tropical and Data Scarce environment., UNESCO-IHE, Institute for Water Education (Ph.D. thesis).

Arrington, D.A., Winemiller, K.O., 2006. Habitat affinity, the seasonal flood pulse, and community assembly in the littoral zone of a Neotropical floodplain river. Journal of the North American Benthological Society 25 (1), 126–141. https://doi.org/10.1899/0887-3593(2006)25[126:hatsfp]2.0.co;2.

Barletta, M., Jaureguizar, A.J., Baigun, C., Fontoura, N.F., Agostinho, A.A., Almeida-Val, V.M.F., Val, A.L., Torres, R.A., Jimenes-Segura, L.F., Giarrizzo, T., Fabré, N.N., Batista, V.S., Lasso, C., Taphorn, D.C., Costa, M.F., Chaves, P.T., Vieira, J.P., Corrêa, M.F.M., 2010. Fish and aquatic habitat conservation in South America: a continental overview with emphasis on neotropical systems. Journal of Fish Biology 76 (9), 2118–2176. https://doi.org/10.1111/j.1095-8649.2010.02684.x.

Barnhill, B., Lopez, E., 1974. Estudio sobre la biologia de los peces del Rio Vinces. Instituto Nacional de Pesca Boletin Cientifico y Tecnico 3 (1), 1–40.

Bovee, K.D., 1982. A Guide to Stream Habitat Analysis Using the Instream Flow Incremental Methodology. IFIP No. 12 FWS/OBS. – edn.

Bovee, K.D., 1986. Development and Evaluation of Habitat Suitability Criteria for Use in the Instream Flow Incremental Methodology. Washington, D.C., 235.

Bovee, K.D., Lamb, B.L., Bartholow, J.M., Stalnaker, C.B., Taylor, J., 1998. Stream Habitat Analysis Using the Instream Flow Incremental Methodology (DTIC Document).

Brown, S.K., Buja, K.R., Jury, S.H., Monaco, M.E., Banner, A., 2000. Habitat suitability index models for eight fish and invertebrate species in casco and sheepscot bays, Maine. North American Journal of Fisheries Management 20 (2), 408–435. https://doi.org/10.1577/1548-8675(2000)020<0408:HSIMFE>2.3.CO;2.

Casatti, L., Mendes, H.F., Ferreira, K.M., 2003. Aquatic macrophytes as feeding site for small fishes in the rosana reservoir, paranapanema river, southeastern Brazil. Brazilian Journal of Biology 63, 213–222.

Costa, R.M.S., Martínez-Capel, F., Muñoz-Mas, R., Alcaraz-Hernández, J.D., Garófano-Gómez, V., 2012. Habitat suitability modelling at mesohabitat scale and effects of dam operation on the endangered júcar nase, parachondrostoma arrigonis (river cabriel, Spain). River Research and Applications 28 (6), 740–752. https://doi.org/10.1002/rra.1598.

Costa, M.R.d., Mattos, T.M., Borges, J.L., Araújo, F.G., 2013. Habitat preferences of common native fishes in a tropical river in Southeastern Brazil. Neotropical Ichthyology 11, 871–880.

DELTARES, 2016. Ecological Knowledge Base, FISH. https://publicwiki.deltares.nl/display/HBTHOME/04+Fish.

Duel, H., van der Lee, G.E., Penning, W.E., Baptist, M.J., 2003. Habitat modelling of rivers and lakes in The Netherlands: an ecosystem approach. Canadian Water Resources Journal 28 (2), 231–247.

Ferreira, A., de Paula, F.R., de Barros Ferraz, S.F., Gerhard, P., Kashiwaqui, E.A.L., Cyrino, J.E.P., Martinelli, L.A., 2012. Riparian coverage affects diets of characids in neotropical streams. Ecology of Freshwater Fish 21 (1), 12–22. https://doi.org/10.1111/j.1600-0633.2011.00518.x.

Florencio, A., 1993. Estudio bioecológico de la laguna Abras de Mantequilla. In: Vinces-Ecuador Revista de Ciencias del Mar y Limnologia, vol. 3. Instituto Nacional de Pesca (INP), pp. 171–192.

Freeman, M.C., Bowen, Z.H., Bovee, K.D., Irwin, E.R., 2001. Flow and habitat effects on juvenile fish abundance in natural and altered flow regimes. Ecological Applications 11 (1), 179–190. https://doi.org/10.2307/3061065.

Funk, A., Winkler, P., Hein, T., Diallo, M., Kone, B., Alvarez-Mieles, G., Pataki, B., Namaalwa, S., Kaggwa, R., Zsuffa, I., D'Haeyer, T., Cools, J., 2013. Balancing Ecology with Human Needs in Wetlands (Fact sheet).

Galecio, E., 2013. Hydrodynamic and Ecohydrological Modelling in a Tropical Wetland: The Abras de Mantequilla Wetland (Ecuador). UNESCO-IHE.

Gery, J., 1977. Characoids of the World. T.F.H. Publications, Inc., Neptune City, NJ, USA, 672 pp.

Glodek, G.S., 1978. The Freshwater Fishes of Western Ecuador. Northern Illinois University, Dekalb, Illinois.

Haasnoot, M., Verkade, J.S., de Bruijn, K.M., 2009. Habitat, a Spatial Analysis Tool for Environmental Impact and Damage Assesment. Deltares, 10.

Halls, A.J. (Ed.), 1997. Wetlands, Biodiversity and the Ramsar Convention: The Role of the Convention on Wetlands in the Conservation and Wise Use of Biodiversity. Ramsar Convention Bureau, Gland, Switzerland.

INP, 2012. Monitoreo de organismos bioacuáticos y recursos pesqueros en el Rio Baba. Preparado para HIDROLITORAL Instituto Nacional de Pesca, Guayaquil, 484.

Jowett, I.G., 1997. Instream flow methods: a comparison of approaches. Regulated Rivers: Research & Management 13 (2), 115–127. https://doi.org/10.1002/(SICI)1099-1646(199703)13:2<115::AID-RRR440>3.0.CO;2–6.

Kerle, F., Zöllner, F., Schneider, M., Kappus, B., Baptist, M.J., 2002. Modelling of long-term fish habitat changes in restored secondary floodplain channels of the River Rhine. In: Paper Presented at the Conference Proceedings of the Fourth International Ecohydraulics Symposium, Cape Town, South Africa.

Laaz, E., Torres, A., 2014. Lista de Peces continentales de la Cuenca del Río Guayas. Available in: http://condordepauledu/waguirre/fishwestec/introhtml.

Laaz, E., Salazar, V., Torres, A., 2009. Guía Ilustrada para la identificación de peces continentales de la Cuenca del Río Guayas (In Spanish). Facultad de Ciencias Naturales – Universidad de Guayaquil, Ecuador.

Loh, M., Vital, W., Vu, V., Navarrete, R., Calle, P., Shervette, V., Torres, A., Aguirre, W., 2014. Isolation of sixteen microsatellite loci for Rhoadsia altipinna (Characiformes: Characidae) from an impacted river basin in western Ecuador. Conservation Genetics Resources 6 (1), 229–231. https://doi.org/10.1007/s12686-013-0062-y.

Maldonado-Ocampo, J.A., Usma Oviedo, J.S., Villa-Navarro, F.A., Ortega-Lara, A., Prada-Pedreros, S., Jimenez, L.F., Jaramillo-Villa, U., Arango, A., Rivas, T.S., Sanchez Garces, G.C., 2012. Peces Dulceacuícolas del Chocó Biogeográfico de Colombia.

McClain, M.E., Subalusky, A.L., Anderson, E.P., Dessu, S.B., Melesse, A.M., Ndomba, P.M., Mtamba, J.O. D., Tamatamah, R.A., Mligo, C., 2014. Comparing flow regime, channel hydraulics, and biological communities to infer flow–ecology relationships in the Mara River of Kenya and Tanzania. Hydrological Sciences Journal 59 (3–4), 801–819. https://doi.org/10.1080/02626667.2013.853121.

Meerhoff, M., Clemente, J.M., De Mello, F.T., Iglesias, C., Pedersen, A.R., Jeppesen, E., 2007. Can warm climate-related structure of littoral predator assemblies weaken the clear water state in shallow lakes? Global Change Biology 13 (9), 1888–1897. https://doi.org/10.1111/j.1365-2486.2007.01408.x.

Meschiatti, A., Arcifa, M., Fenerich-Verani, N., 2000. Fish communities associated with macrophytes in Brazilian floodplain lakes. Environmental Biology of Fishes 58 (2), 133–143. https://doi.org/10.1023/a:1007637631663.

Milhous, R., Waddle, T., 2012. Physical Habitat Simulation (PHABSIM) software for windows (v. 1.5. 1). In: Fort Collins, CO: USGS Fort Collins Science Center Institute of Freshwater, Research Drottningholm, Sweden.

Mitsch, W.J., Gosselink, J.G., 2007. Wetlands, fourth ed. John Wiley & Sons, Inc, Hoboken, New Jersey, USA.

Mouton, A.M., Alcaraz-Hernández, J.D., De Baets, B., Goethals, P.L.M., Martínez-Capel, F., 2011. Data-driven fuzzy habitat suitability models for brown trout in Spanish Mediterranean rivers. Environmental Modelling & Software 26 (5), 615–622. https://doi.org/10.1016/j.envsoft.2010.12.001.

Muñoz-Mas, R., Martínez-Capel, F., Schneider, M., Mouton, A.M., 2012. Assessment of brown trout habitat suitability in the Jucar River Basin (Spain): comparison of data-driven approaches with fuzzy-logic models and univariate suitability curves. The Science of the Total Environment 440, 123–131. https://doi.org/10.1016/j.scitotenv.2012.07.074.

Prado, M., 2009. Aspectos biologicos y pesqueros de los principales peces de aguas continentales de la provincia de Los Ríos durante julio de 2009. Boletín Instiuto Nacional de Pesca, Ecuador, pp. 11–13. No 5.

Prado, M., Revelo, W., Castro, R., Bucheli, R., Calderón, G., Macías, P., 2012. Caracterización química y biológica de sistemas hídricos en la Provincia de Los Ríos-Ecuador (in Spanish). In: vol Boletín Científico y Técnico, vol. 20. Instituto Nacional de Pesca, Guayaquil, Ecuador, pp. 1–100, 100.

Quevedo, O., 2008. Ficha Ramsar del Humedal Abras de Mantequilla - Ecuador (In Spanish) (Guayaquil, Ecuador).

Ramsar, 2013. The Ramsar Convention Manual: A Guide to the Convention on Wetlands (Ramsar, Iran, 1971), sixth ed. Ramsar Convention Secretariat Ramsar Convention Secretariat, Gland, Switzerland.

Ramsar, 2014. The List of Wetlands of International Importance. In: RAMSAR. http://www.ramsar.org/pdf/sitelist.pdf.

Revelo, W., 2010. Aspectos Biológicos y Pesqueros de los principales peces del Sistema Hídrico de la Provincia de Los Ríos, durante 2009. In: Boletin Cientifico y Técnico, vol 20. Instituto Nacional de Pesca, Guayaquil-Ecuador, pp. 53–84.

Roberts, T.R., 1973. The Glandulocaudine Characid Fishes of the Guayas Basin in Western Ecuador, vol. 144. Bulletin of The Museum of Comparative Zoology. Available in: http://www.biodiversitylibrary.org/page/4228014#page/519/mode/1up.

Schamberger, M.A., Farmer, A.H., Terrell, J.W., 1982. Habitat Suitability Index Models: Introduction. U.S. D.l. Fish and Wildlife Service. FWS/OBS-82/10. 2 pp.

Schneider, M., Noack, M., Gebler, T., Kopecki, L., 2010. Handbook for the Habitat Simulation Model CASiMiR. Module CASiMiR-fish. Base Version. SJE–Schneider & Jorde Ecological Engineering GmbH LWW–Institut fur Wasserbau, Universitat Stuttgart.

Schneider, M., Kopecki, I., Eberstaller, J., Frangez, C., Tuhtan, J.A., 2012. Application of CASiMiR-GIS for the simulation of brown trout habitat during rapid flow changes. In: Paper Presented at the Proceedings of the 9th International Symposium on Ecohydraulics (ISE 2012).

Spence, R., Hickley, P., 2000. The use of PHABSIM in the management of water resources and fisheries in England and Wales. Ecological Engineering 16 (1), 153–158. https://doi.org/10.1016/S0925-8574(00)00099-9.

Teixeira-de Mello, F., Meerhoff, M., Pekcan-Hekim, Z., Jeppesen, E., 2009. Substantial differences in littoral fish community structure and dynamics in subtropical and temperate shallow lakes. Freshwater Biology 54 (6), 1202–1215. https://doi.org/10.1111/j.1365-2427.2009.02167.x.

Teresa, F.B., Casatti, L., 2013. Development of habitat suitability criteria for Neotropical stream fishes and an assessment of their transferability to streams with different conservation status. Neotropical Ichthyology 11, 395–402.

Tuhtan, J.A., Wieprecht, S., 2012. No going back: including second law irreversibility in fish habitat models. In: Paper Presented at the 9th ISE 2012, Vienna.

Winemiller, K.O., Jepsen, D.B., 1998. Effects of seasonality and fish movement on tropical river food webs. Journal of Fish Biology 53, 267–296. https://doi.org/10.1111/j.1095-8649.1998.tb01032.x.

Wolfshaar, K.E. van de, Ruizeveld de Winter, A.C., Straatsma, M.W., Brink, N.G.M., Leeuw, J.J. de, 2010. Estimating spawning habitat availability in flooded areas of the river Waal, The Netherlands. River Research and Applications 26 (4), 487–498.

Zorn, T.G., Seelbach, P.W., Rutherford, E.S., 2012. A regional-scale habitat suitability model to assess the effects of flow reduction on fish assemblages in Michigan Streams1. JAWRA: Journal of the American Water Resources Association 48 (5), 871–895. https://doi.org/10.1111/j.1752-1688.2012.00656.x.

7

A Comparison of Spatial–Temporal Scale Between Multiscalar Drought Indices in the South Central Region of Vietnam

Hung Manh Le[1], Gerald Corzo[2], Vicente Medina[3], Vitali Diaz[2,4], Bang Luong Nguyen[5], Dimitri P. Solomatine[2,4]

[1]NATIONAL CENTRAL FOR WATER RESOURCES PLANNING AND INVESTIGATION (NAWAPI), MINISTRY OF NATURAL RESOURCES AND ENVIRONMENT (MONRE) OF VIETNAM, HANOI, VIETNAM; [2]UNESCO-IHE DELFT INSTITUTE OF WATER EDUCATION, DELFT, THE NETHERLANDS; [3]THERMAL ENGINES DEPARTMENT, TECHNICAL UNIVERSITY OF CATALONIA, BARCELONA, SPAIN; [4]WATER RESOURCES SECTION, DELFT UNIVERSITY OF TECHNOLOGY, DELFT, THE NETHERLANDS; [5]FACULTY OF WATER RESOURCES ENGINEERING, THUY LOI UNIVERSITY, HANOI, VIETNAM

1. Highlights

- Comparison of spatial–temporal drought characteristics between the Standardized Precipitation Index (SPI) and Standardized Precipitation Evapotranspiration Index (SPEI).
- Drought severity classified using Non-Contiguous Drought Area (NCDA).
- Important role of temperature detected in terms of drought magnitude.

2. Introduction

It is widely recognized that drought is an inevitable and recurrent natural hazard. People are more vulnerable to drought than any other hazard due to its complexity and lack of understanding (Wilhite, 2000). Between 1970 and 2014, droughts threatened the lives of 1.62 billion people in Asia and the Pacific and caused economic losses of 53 billion US dollars (UNESCAP, 2015). Prolonged drought is associated with persistent unemployment, widespread uncultivated land, which often leads to migration as a coping strategy, and increase in the risk of conflict due to the decline in natural

Spatiotemporal Analysis of Extreme Hydrological Events. https://doi.org/10.1016/B978-0-12-811689-0.00007-0

resources (IOM, 2014). For example, there is evidence that the most severe drought during 2007–10 played an important role in the endless civil war in Syria. The drought caused massive failure in agriculture and livestock deaths. The unexpected consequence of 1.5 million people migrating from rural farming areas to urban areas was believed to be the cause of the onset of the conflict (Kelley et al., 2015).

Drought assessment considers several aspects such as its duration, intensity, spatial extent, and socioeconomic impact in the affected area. Wilhite and Glantz (1985) classified the definition of drought in four types: meteorological drought, agricultural drought, hydrological drought, and socioeconomic drought. If rainfall deficiency and high temperatures (meteorological drought) last long enough, they will trigger a shortage of soil moisture, posing a threat to the water demands of plants (agricultural drought). Soil moisture deficiency also causes a reduction in streamflow: inflow to reservoirs, lakes, and ponds (hydrological drought). With a prolonged drought, socioeconomic impacts may happen as negative consequences resulting from agricultural and hydrological drought.

Meteorological drought is normally triggered first and consequently leads to other types of drought. As a result, monitoring meteorological drought can be beneficial when considering the degree of severity of the whole drought process. Therefore this study will mainly focus on meteorological drought indices. Another advantage of meteorological drought indices is that they often require fewer input variables such as precipitation or temperature, which makes them applicable in many regions. It is noticed that there exist a number of satellite-based indices but these are beyond the content of this study. We focus on ground-based indices since observed ground rainfall and temperature are available for the South Central Region of Vietnam.

Some of the earliest indices were proposed based on precipitation, e.g., Munger's index (Munger, 1916) and Blumenstock's index (Blumenstock, 1942). A significant advancement was made by Palmer (1965) when the author used the principle of water balance to develop the so-called Palmer Drought Severity Index (PDSI). This index has been widely used for many years, especially for monitoring drought in the United States. A shortcoming of the PDSI lies in its empirical parameters estimated from typical climates of several stations in the United States, which may not be true on a global scale. To overcome this problem, Wells et al. (2004) inferred an update to this index (called the self-calibrating PDSI) and applied it to different climate systems. Yet, an absence of duration time and slowness in detecting the onset of drought still remain. The response time of different components of the hydrological system (e.g., surface water, ground water) to precipitation anomalies varies, thus a multitime scale drought index is needed. McKee et al. (1993) clearly illustrated this characteristic of droughts and developed a precipitation-based index (SPI). Having different time scales, which range from 1 to 48 months, and clear mathematic inferences are the advantages of the SPI. An important feature of the SPI (explicated by its name) is that this index transforms accumulated precipitation anomalies' time series to standard normalized distribution values. With that, we can compare drought conditions in any climate region. Moreover, in contrast to the PDSI, the SPI needs only precipitation as an input variable, which enables regions with limited data to apply it. The SPI has been studied in various drought aspects such as frequency analysis (Mishra and Singh, 2009), the response of hydrological systems to temporal drought (Vicente-Serrano and López-Moreno, 2005), and spatial–temporal drought (Bayissa et al., 2015; Dahal et al., 2016). In 2009, the World Meteorological Organization announced the Lincoln Declaration on Drought Indices, which agreed on the use of the SPI as a universal meteorological index for more effective drought monitoring and climate risk management (WMO, 2009). It is recommended that all National Meteorological and Hydrological Services around the world should include the SPI in their services.

On the other hand, there exist several potential weaknesses concerning the application of the SPI. McKee et al. (1993) proposed a two-parameter gamma distribution function to fit the rainfall time series, but the use of this distribution is debatable. For example, Stagge et al. (2015) agreed that gamma distribution is recommended for use in European territories, while Quiring (2009) and Vicente-Serrano (2006) suggested a three-parameter Pearson Type III as a more appreciable function. Moreover, the accuracy of the SPI is associated with the length of precipitation time series (Mishra and Singh, 2010). Ideally, data of more than 30 years can produce more reliable drought information (McKee et al., 1993; WMO, 2012). Similar to other precipitation-based techniques, the SPI assumes that precipitation primarily dominates drought conditions and there is no trend for other climatic variables (e.g., temperature). This assumption is a flaw for regions that are highly vulnerable to climate change. In addition, the lack of atmospheric water demand could allow the SPI to fail to detect historical droughts in the dry climate areas that experience many months without precipitation (McEvoy et al., 2012). Focusing on this limitation, Vicente-Serrano et al. (2010) proposed a new index, which includes a temperature variable, namely, the SPEI. Basically, the SPEI calculation framework is the same as SPI's. The main difference between SPI and SPEI lies in the time series used to fit into a distribution function are accumulated climatic water balance anomalies, defined as a deficit between precipitation and potential evapotranspiration. The usefulness of the SPEI in drought assessment has been proved in various places across the world, such as Australia (Deo and Şahin, 2015), Czech Republic (Potop et al., 2012), Jordan (Törnros and Menzel, 2014), Spain (Lorenzo-Lacruz et al., 2010), and the United States (McEvoy et al., 2012). The better performance of the SPEI over SPI is reported in some works (McEvoy et al., 2012; Vicente-Serrano et al., 2012). The limitation of the SPEI is that it requires more data than just precipitation, and is sensitive to the method of calculating potential evapotranspiration. Additionally, as with other drought indices, a long enough period (e.g., 30–50+ years) to sample natural variability should be employed (Vicente-Serrano and National Center for Atmospheric Research Staff, 2015).

Spatial–temporal drought analysis has been investigated using various approaches. Mishra and Desai (2005) developed the drought severity–area–frequency curve using estimated grid SPIs to assess the severity of the localized drought within the Kangsabati River basin, India. Clustering algorithms have also been attempted in some works. For example, Santos et al. (2010) applied a principal component analysis and K-mean clustering to the SPI series for spatial and temporal patterns of drought in Portugal, while Corzo Perez et al. (2011) used the NCDA and the Contiguous Drought Area approach to characterize drought evolution at a global scale during the period 1963–2000.

In Vietnam, Vu-Thanh et al. (2014) reported that there have been very few assessable papers related to meteorological drought application. One of the significant contributions is the work of Nguyen (2005). The author applied the SPI to 13 provinces in the Central Highlands and South Central Region of Vietnam. It has been concluded that the SPI typically captures the most severe drought events but not moderate ones. By comparing different meteorological drought indices, Vu-Thanh et al. (2014) concluded that there is no particular drought index that can generally characterize drought events in Vietnam, therefore the challenge for a universal meteorological index remains in Vietnam. In 2015, Nguyen et al. (2015) proposed a comprehensive index from a combination of meteorological and hydrological drought indices for the Cai River basin. The integrated index has been evaluated to have better performance when compared to an individual index. Vu et al. (2015) compared observation rain gauges with a number of rainfall sources from global datasets to simulate the SPI on the Central Highlands during the period 1990–2005. The result indicated that the drought index generated by the global dataset can reflect drought events in a given study area.

The primary objective of this study is to evaluate which of the common meteorological indices, SPI or SPEI, is suitable for monitoring drought conditions for the South Central Region of Vietnam. To do this, the NCDA was applied to exploit the evolution of spatial–temporal drought based on the aforementioned indices. Multiple time scales of 3, 6, 9, and 12 months were considered for each drought index. Generally, a short time scale drought index could monitor agricultural drought, while medium to long time scale indices could be useful for hydrological drought.

3. Materials and Methods

3.1 Study Area

The South Central Region of Vietnam covers an area of 27,500.3 km^2, spreading from 10°50′N to 14°50′N and 107°50′E to 109°20′E, including five provinces: Binhdinh, Phuyen, Khanhhoa, Ninhthuan, and Binhthuan. There is 55.2% labor in this region working in agriculture, forestry, and fishing, much higher than the national average (44.7%; General Statistic Office of Vietnam, 2014). Global Land Cover 300 m resolution v1.6.1 for the year of 2010 in the Climate Research Data Package was used to explore land use in the study (Fig. 7.1). It is clear from the figure that cropland is a primary land use, making up 45.2% of total land, followed by forest land and shrubland with total proportions of 26.5% and 20.8%, respectively (Fig. 7.1). Of those croplands, rainfed fields are the most dominant type, because agricultural production in the study area depends heavily on natural

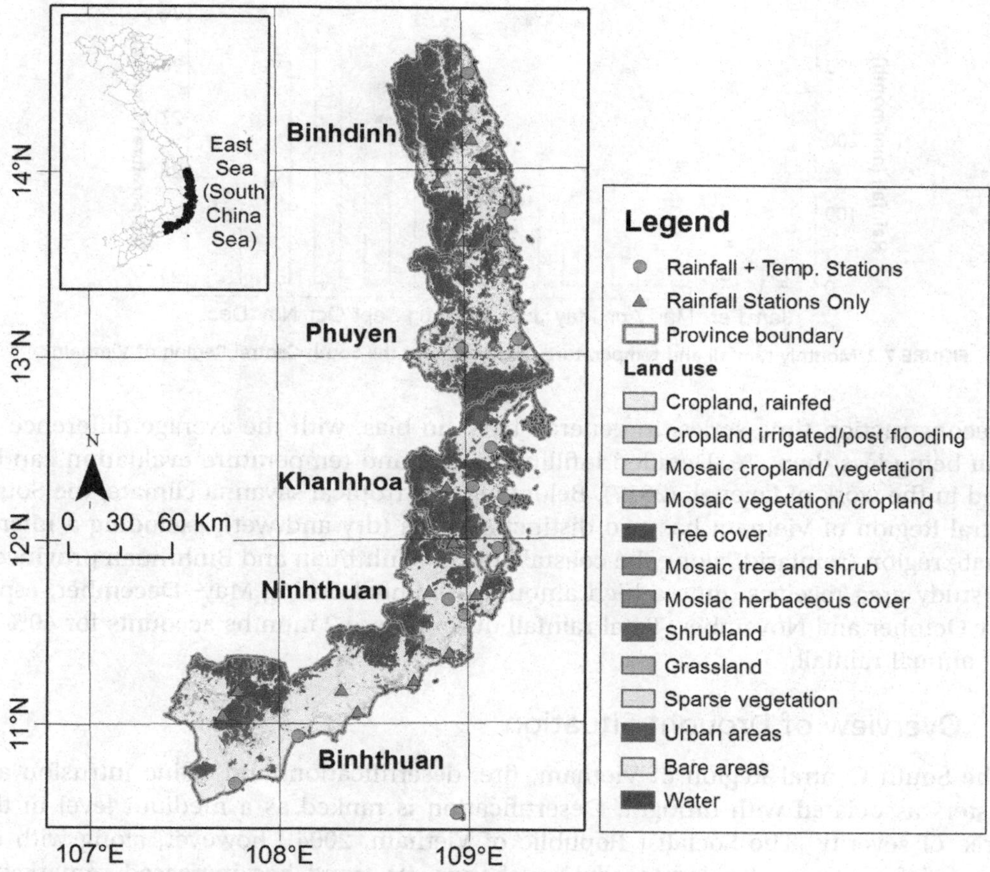

FIGURE 7.1 Land use and land cover of the South Central Region of Vietnam and meteorological stations distribution map.

variation. Paddy rice and maize are the main cereal crops in the South Central Region of Vietnam. Several perennial plants are intensively tilled such as sugarcane, peanut, and coconut. Livestock rearing is widespread in the region depending on the topography and climate. Buffalo and cow can be found in most of the lowland areas, while sheep are dominant for the dry climate conditions in Ninhthuan and Binhthuan provinces.

3.2 Data

This study collected data from 30 monthly rainfall stations and 13 monthly temperature stations. Distribution of monthly rainfall and temperature for all stations is presented in Fig. 7.2. Also, descriptions of those stations are presented in Appendices 1 and 2. The data length of each station is a period of 38 years (1977–2014), but several of them have reconstructed data. For the reconstructed data, the postreconstruction time series and

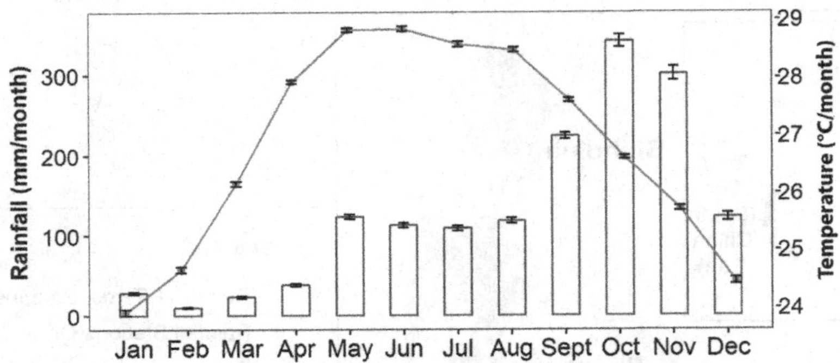

FIGURE 7.2 Monthly rainfall and temperature distribution of the South Central Region of Vietnam.

prereconstruction time series, in general, have no bias, with the average difference in mean being less than 5%. Detailed infilling rainfall and temperature evaluation can be found in the work of Le et al. (2017). Belonging to a tropical savanna climate, the South Central Region of Vietnam has two distinct seasons (dry and wet), excluding a microclimate region (semiarid) along the coastal area of Ninhthuan and Binhthuan provinces. The study area receives quite a high amount of rainfall during May–December, especially October and November. Total rainfall during those 2 months accounts for 40% of total annual rainfall.

3.3 Overview of Drought Situation

In the South Central Region of Vietnam, fire, desertification, and saline intrusion are disasters associated with drought. Desertification is ranked as a medium level in the degree of severity (The Socialist Republic of Vietnam, 2004); however, along with an increased frequency of extreme climate change, its trend has increased remarkably (Pham and Le, 2009). This disaster seriously affects provinces located in the southern part of the region, which are Khanhhoa, Ninhthuan, and Binhthuan, with a total area of 300,000 ha (Gobin et al., 2012).

The study area has been prone to droughts for many years, in addition to being vulnerable to floods. A large portion of drought often occurs during the summer–autumn season (July–September) (Dao, 2002; Shaw et al., 2007). Based on the affected area in drought, drought duration and intensity, three extreme cross-year droughts during the given study period are identified as 1982–83, 1997–98, and 2004–05 events. The prolonged drought from 1982 to 1983 destroyed a total of 291,000 ha of crops in central and southern Vietnam, while the drought in 1998 caused a total drying of the crop, reaching 20.3%–25% of planting area. Rainfall was much lower in the winter–spring of 1997–98 (November–March), as compared with the average rainfall, and continued to drop until summer (Nguyen and Shaw, 2011). As a consequence, widespread fresh water shortage affected 203,000 people in the South Central Region (Nguyen, 2005). A drought assessment by the Ministry of Agriculture and Rural

Development for the whole of Vietnam concluded that this extreme drought period affected 3 million people and total economic losses in terms of agricultural production were estimated to be about 400 million US dollars. In 2004, drought mainly affected Ninhthuan province, with normal rainfall declining by half, and expanding to the whole South Central Region in the following year (Nguyen and Shaw, 2011). The major drought of 2004–05 in Ninhthuan province significantly impacted on agriculture, forestry, and aquatic production. There were 35,276 households using emergency food aid, more than 55,000 cows and 36,000 goats lacking water and food, and total economic losses up to April 2005 were estimated at 137 billion VND. Likewise, Binhthuan province was also reported to be struggling with a major drought, with a famine relief campaign for 76,000 people and about 221 billion VND economic loss (Pham and Le, 2009). These huge losses contributed in part to the lack of preparedness and awareness of regional governments and communities (Shaw et al., 2007). The exceptionally strong El Niño in 2015–16 was believed to have affected the absence of rainfall in a large area of Vietnam for a long period, typically Vietnam's South Central Region. For example, at Khanhhoa province, 10,000 ha of agricultural land stopped producing crops (MARD, 2016). Tran (2016) conducted a survey of 250 people in Ninhthuan province in terms of usable water resources. A total of 129/250 respondents reported they were always lacking water for living, and 242/250 respondents said that they always lacked water for crop production. Leaving land for cultivation due to drought is a serious problem; 95.2% of respondents found themselves in this position and almost all the survey answers considered that frequent drought made them increasingly poorer.

3.4 Calculation of Drought Indices

Three main steps used to calculate the SPI and SPEI are presented in Table 7.1. Generally, the SPEI's calculation steps are the same as SPI's. The SPI and SPEI are calculated on multiple time scales. Those time scales have the benefit of reflecting the availability of different water resources affected by drought. The SPI is calculated on a monthly time scale, while the SPEI quantifies drought on a weekly basis. This study only works with monthly data so the calculation methodology focuses on the monthly scale. The aim of calculation is that the desired time series is fitted to a probability distribution, which is then transformed into a normal distribution so that the mean of the SPI (or SPEI) is equal to zero in the given period. Because the SPI and SPEI are normalized, dryness or wetness is represented in the same manner. Therefore wet periods can be monitored by the SPI and SPEI as well.

The SPI or SPEI is classified into different categories as shown in Table 7.2. McKee et al. (1993) determined that the SPI indicates moderate drought 9.2% of the time, severe drought 4.4% of the time, and extreme drought 2.3% of the time. This is equivalent to a 10-, 20-, and 50-year return period for moderate drought, severe drought, and extreme drought, respectively. Due to the similar manner of calculation, the SPEI also inherits those features from the SPI.

Table 7.1 Three Steps to Calculate the Standardized Precipitation Index/Standardized Precipitation Evapotranspiration Index (SPI/SPEI)

SPI	SPEI
(1) Fitting Distribution Function for Time Series X	
Establishment of time series X for a different time scale k (monthly)	
–	$D_{i,m} = P_{i,m} - \mathrm{PET}_{i,m}$ (7.1)
$X_{i,m}^k = \sum_{t=13-k+m}^{12} P_{i-1,t} + \sum_{t=1}^{m} P_{i,t}$ (7.2)	$X_{i,m}^k = \sum_{t=13-k+m}^{12} D_{i-1,t} + \sum_{t=1}^{m} D_{i,t}$ (7.3)
where k is the time scale, i is the year, m is the month in year, P is the precipitation, and PET[a] is potential evapotranspiration.	
The following are examples of 1-month and 3-month scale calculations in the year 1989: $X_{1989,1}^1 = P_{1989,1}$ $X_{1989,1}^3 = P_{1988,11} + P_{1988,12} + P_{1989,1}$	$X_{1989,1}^1 = D_{1989,1}$ $X_{1989,1}^3 = D_{1988,11} + D_{1988,12} + D_{1989,1}$
Fitting distribution for each month, final result is a total of 12 fitting functions corresponding to 12 months	
Fitting function: two-parameter gamma distribution[b] Cumulative probability distribution function $G(x) = \int_0^x x^{\alpha-1} e^{-x/\beta} dx$ (7.4) where α is the shape parameter and β is the scale parameter.	Fitting function: three-parameter log-logistic distribution[b] Cumulative probability distribution function $F(x) = \left[1 + \left(\frac{\alpha}{x-\gamma}\right)^{\beta}\right]^{-1}$ (7.5) where α is the shape parameter, β is the scale parameter, and γ is the original parameter.

(2) The Cumulative Probability of Time Series X Is Computed Relative to the Fitting Distribution

$H(x) = p + (1 - p)G(x)$ (7.6) where p is the probability of no precipitation.	$H(x) = F(x)$ (7.7)

(3) The Cumulative Probability Is Transformed to the Standard Normal Variable and the SPI or SPEI Is Found

$$Z = \begin{cases} -\left(W - \dfrac{C_0 + C_1W + C_2W^2}{1 + d_2W + d_2W^2 + d_3W^3}\right) & 0 < H(x) \le 0.5 \\ +\left(W - \dfrac{C_0 + C_1W + C_2W^2}{1 + d_2W + d_2W^2 + d_3W^3}\right) & 0.5 < H(x) \le 1 \end{cases} \quad (7.8)$$

where:

$$Z = \begin{cases} \sqrt{-2\ln(H(x))} & 0 < H(x) \le 0.5 \\ \sqrt{-2\ln(1 - H(x))} & 0.5 < H(x) \le 1 \end{cases}$$

and $C_0 = 2.515517, C_1 = 0.802853, C_2 = 0.010328$
$d_1 = 1.432788, d_2 = 0.189269, d_3 = 0.001308$
and Z is represented for the SPI or SPEI.

[a]In this study, the Thornthwaite method was used to calculate potential evapotranspiration.
[b]Functions were suggested by authors who developed those indices.

Table 7.2 Standardized Precipitation Index/Standardized Precipitation Evapotranspiration Index (SPI/SPEI) Classification

Category	SPI/SPEI	Probability of Events (%)	Cumulative Probability
Extreme wet	SPI/SPEI ≥ 2.0	2.3	0.977–1.000
Severe wet	1.5 ≤ SPI/SPEI < 2.0	4.4	0.933–0.977
Moderate wet	1.0 ≤ SPI/SPEI < 1.5	9.2	0.841–0.933
Normal	−1.0 ≤ SPI/SPEI < 1.0	68.2	0.159–0.841
Moderate dry	−1.5 ≤ SPI/SPEI < −1.0	9.2	0.067–0.159
Severe dry	−2.0 ≤ SPI/SPEI < −1.5	4.4	0.023–0.067
Extreme dry	SPI/SPEI ≤ −2.0	2.3	0.000–0.023

3.5 Spatial–Temporal Drought Analysis

Drought indices in the previous section only reflect on a temporal scale; however, the spatial scale is also an important factor to assess the degree of drought severity. Therefore this study developed a methodology using the NCDA approach. In the work of Corzo Perez et al. (2011), the NCDA method was applied for hydrological drought based on binary expression (i.e., 0 for a nondrought event and 1 for a drought event). The SPI or SPEI, on the other hand, can identify various types of drought such as moderate, severe, and extreme events. For that reason, this study proposes an expression such as Eq. (7.9):

$$D_{s,t} = \begin{cases} 0 & \mathrm{SPI}_t/\mathrm{SPEI}_t > -1.0 \\ 1 & -1.0 > \mathrm{SPI}_t/\mathrm{SPEI}_t \geq -1.5 \\ 2 & -1.5 > \mathrm{SPI}_t/\mathrm{SPEI}_t \geq -2.0 \\ 3 & \mathrm{SPI}_t/\mathrm{SPEI}_t < -2.0 \end{cases} \tag{7.9}$$

where $D_{s,t}$ is a drought state per cell at time t, which is determined by the value of the SPI/SPEI at the same time; the values of 1, 2, and 3 are equivalent to moderate, severe, and extreme drought, respectively. The slash indicates the use of either the SPI or SPEI.

Percentage of the area in different types of drought is calculated by following formulas:

$$\mathrm{PDA}_{\mathrm{moderate},t} = \frac{\sum(D_{s,t} = 1) \times A}{A_{\mathrm{tot}}} \times 100 \tag{7.10}$$

$$\mathrm{PDA}_{\mathrm{severe},t} = \frac{\sum(D_{s,t} = 2) \times A}{A_{\mathrm{tot}}} \times 100 \tag{7.11}$$

$$\mathrm{PDA}_{\mathrm{extreme},t} = \frac{\sum(D_{s,t} = 3) \times A}{A_{\mathrm{tot}}} \times 100 \tag{7.12}$$

$$\mathrm{PDA}_{\mathrm{all},t} = \mathrm{PDA}_{\mathrm{moderate},t} + \mathrm{PDA}_{\mathrm{severe},t} + \mathrm{PDA}_{\mathrm{extreme},t} \tag{7.13}$$

where $PDA_{moderate,t}$, $PDA_{severe,t}$, $PDA_{extreme,t}$, and $PDA_{all,t}$ are percentages of areas in moderate, severe, extreme, and all types of drought condition at time t, respectively, A_{tot} is total land area, and A is the grid's cell area.

To undertake an assessment on a spatial scale in the South Central Region of Vietnam, a grid cell with a resolution of 4×4 km was created for the whole region, with a total of 1680 cells. The reason for selecting this resolution size was that this study collected meteorological data from Phuquy Island, with an area of 16.4 km^2. Therefore a grid size of 16 km^2 was enough to represent this island. To calculate drought indices for each cell, the method in the paper by Rhee and Carbone (2011) to produce better spatial interpolation prior to drought indices calculation was applied. Rainfall and temperature from measurement sites were first interpolated at each grid cell using the inverse distance weighting (IDW) method, and then the SPI or SPEI was calculated for each cell based on interpolated rainfall and temperature. Le et al. (2017) pointed out that IDW is less accurate than advanced methods (multiple linear regression, artificial neural network) for interpolating rainfall and temperature. However, there is no clear difference between IDW and other methods in terms of drought event estimation. Therefore, IDW has been applied to set up the rainfall and temperature grids due to its ease of use.

4. Results

4.1 Distribution Testing of the SPI and SPEI

Transforming from highly skewed accumulated rainfall anomalies to standardized normal distribution (SPI) or accumulated climate water balance anomalies to standardized normal distribution (SPEI) needs a proper distribution function. An unappreciated selection can induce a bias in drought indices calculation, which is an overestimate or underestimate of the dryness or wetness situation (Sienz et al., 2012). Gamma distribution and log-logistic distribution have been suggested for the SPI and SPEI respectively by their developers and were used in this study. Figs. 7.3 and 7.4 present distribution maps of the standard deviations of fitting time series and parameters for each of the fitting functions. It is observed that rainfall or water balance variance is larger on the northern part than on the southern part. A similar pattern was found for most of the parameters, except for the scale parameter of the log-logistic distribution function. Regarding this distribution, with the cell having a small standard deviation of water balance time series, its shape parameter tends to be smaller as well. The scale parameter of the log-logistic distribution function is negative for all grid cells, while that of the gamma distribution, in turn, has all positive values.

To validate the fitting quality of each time series, we performed the goodness of a fit test using a two-sample Kolmogorov–Smirnov test (K–S test) (Darling, 1957). The null hypothesis of the K–S test is that both distributions follow the same distribution. A P-value less than 0.05 means a rejection of this null hypothesis. The advantage of the

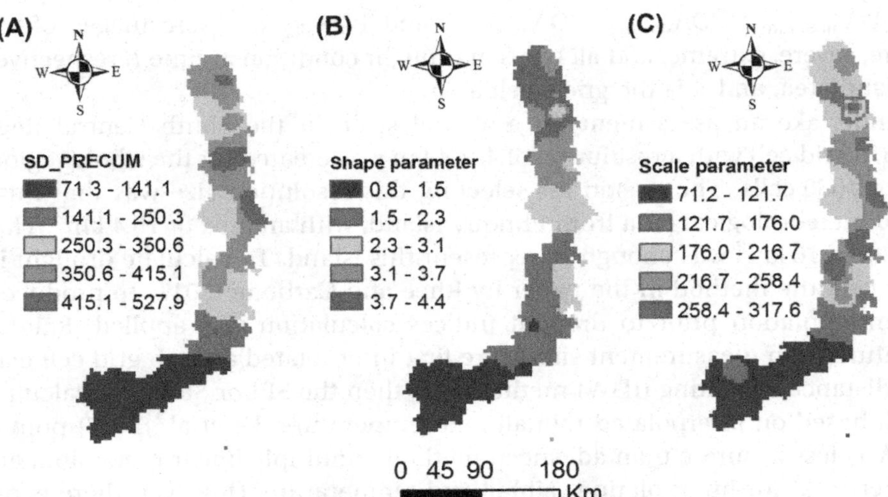

FIGURE 7.3 Distribution of (A) standard deviation 3-month cumulative precipitation, (B) shape parameter, and (C) scale parameter of gamma function for SPI3.

K–S test for the fitting model is that this test checks the fit for the entire dataset (including extreme values) and not just the traditional two parameters (mean and variance) (Sager, 2010). It also reveals the part in which theoretical cumulative distribution function (CDF) departs from empirical CDF. SPI or SPEI calculation gives us a total of 12 different fitting functions for each grid cell. Fig. 7.5 shows the cumulative percentage of the K–S test's *P*-value for drought indices. Except for about 1% of grid cells at 3 months, the SPI rejected the null hypothesis; the rest proved that rainfall in

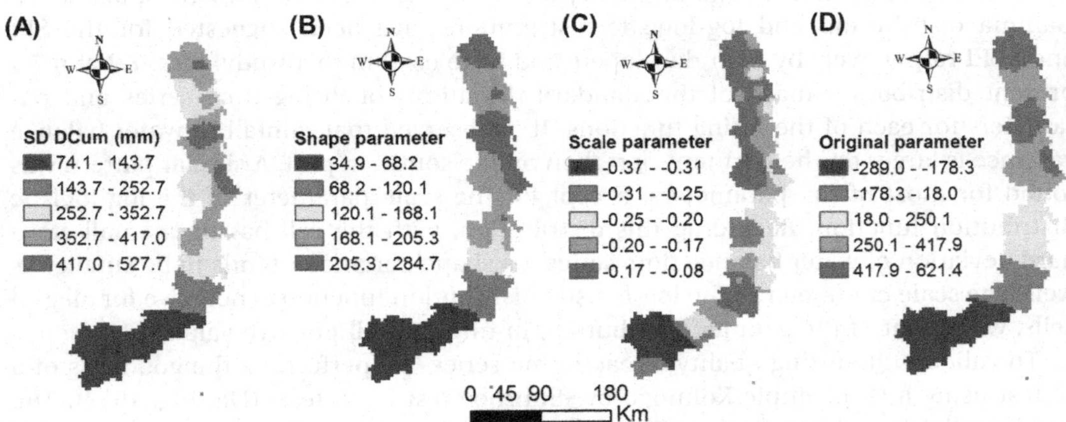

FIGURE 7.4 Distribution of (A) standard deviation 3-month cumulative precipitation minus potential evapotranspiration, (B) shape parameter, (C) scale parameter, and (D) original parameter of log-logistics function for SPEI3.

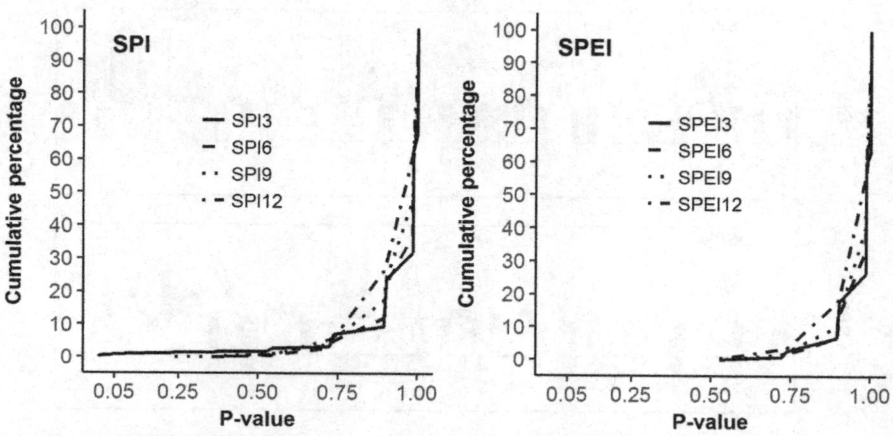

FIGURE 7.5 Cumulative percentage of *P*-value of Kolmogorov–Smirnov test test for Standardized Precipitation Index (SPI) fitting (*left*) and Standardized Precipitation Evapotranspiration Index (SPEI) fitting (*right*).

the study area and water balance (rainfall minus potential evapotranspiration) follow gamma distribution and log-logistic distribution, respectively. The next section discusses further implications of both indices on spatial–temporal assessment.

4.2 Temporal Drought Evolution in the South Central Region of Vietnam

The SPI and SPEI over the South Central Region of Vietnam were calculated by taking the average of corresponding gridded SPIs and SPEIs. Figs. 7.6 and 7.7 present the temporal drought at different time scales of those drought indices' time series during the period 1977–2014. Overall, the SPI revealed that the 1980 and 1990s were dry years, while 1980–82, 1999–2001, and 2008–10 were wet years. Starting from 2011, the study area started to suffer prolonged drought periods. Similarly, the SPEI time series also gave the same dry and wet year patterns as the SPI, but it seems the severity of the drought of 2011–14 was more extreme. Additionally, during the period 1997–98, the SPEI time series showed an extreme drought period but it was not clearly observed for the SPI time series.

4.3 Spatial Drought Assessment (NCDA Method)

By analyzing past drought events, Pham and Le (2009) concluded that currently droughts in the South Central Region of Vietnam are associated with the absence of rainfall from the prior year. For this reason, it is better to consider 2 consecutive years' drought instead of a calendar year's. It is noticed that US drought monitoring uses the concept of the hydrological year (prior year October to current year September) to classify the drought situation, and therefore this study also adopted this notion. As mentioned in Section 3.3, the study area suffered three prolonged drought events in 1982–83,

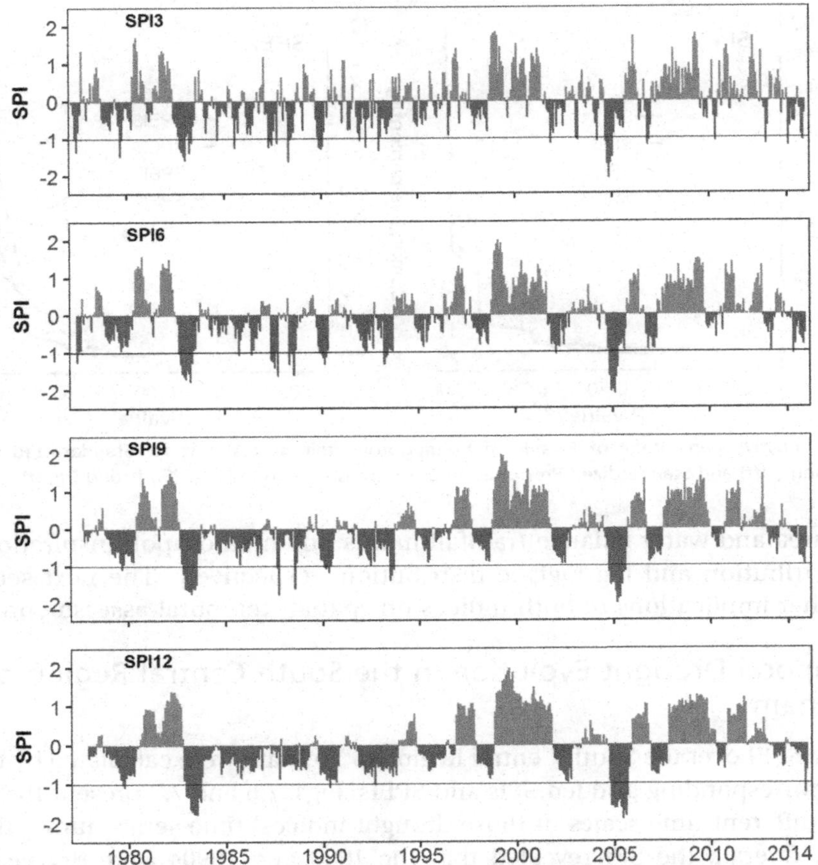

FIGURE 7.6 Evolution of temporal drought at multiple Standardized Precipitation Index (SPI) time scales in the South Central Region of Vietnam. X axis ticks are the first month of the year. The *solid black line* is drought threshold identified by the SPI.

1997–98, and 2004–05. Therefore we propose as a condition for the suitability of the drought index for the South Central Region of Vietnam that the drought index at least can capture all three events (1982–83, 1997–98, and 2004–05).

The percentage drought in area (PDA) approach was applied for a spatial temporal assessment. PDA_{all} at different time scales based on the SPI and SPEI is exploited in Fig. 7.8. Blue indicates the 1982–83 period, red indicates the 1997–98 period, and 2004–05 is represented in green, while the other periods are displayed in gray. It is clear that in different time scales of both the SPI and SPEI, the PDA during 1982–83 and 2004–05 was much higher than others. The development of drought conditions during those periods had a similar style. In a short time scale (3 months), it was detected that drought was the most severe during the middle of the prior year to January of the current year. Typically, PDA_{all} of SPEI3 at January 1983 was approximately 100%, which

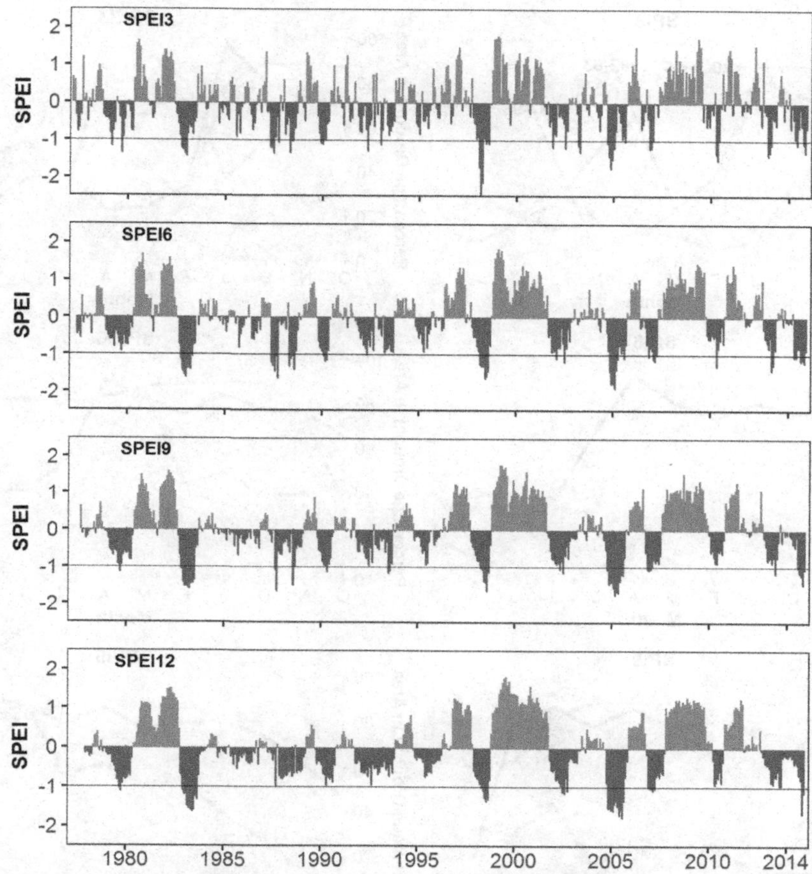

FIGURE 7.7 Evolution of temporal drought at multiple Standardized Precipitation Evapotranspiration Index (SPEI) time scales in the South Central Region of Vietnam. X axis ticks are the first month of the year. The *solid black line* is drought threshold identified by the SPEI.

indicated almost 100% of the study area was in drought condition from November 1982 to January 1983. There was another short drought event detected by SPI3 for the period February 1983–April 1983, with total percentage drought in the area in those months of over 60%. There was a shifted time in an assessment of drought magnitude with longer time scales. For example, in the medium time scale (6 months), the severity of drought was extended to April 1982 or March 2005, while 9- and 12-month time scales suggested an impact until June–September. The timing of drought evolution of short and medium time scales for 1982–83 and 2004–05 is quite similar to agriculture reports on drought damage.

Interestingly, the SPI underestimated the drought situation in 1997–98, indicated by the fact that SPI's PDA_{all} for this period was mixed with those of other years at all time scales. On the contrary, this period was marked as a serious drought episode by the SPEI.

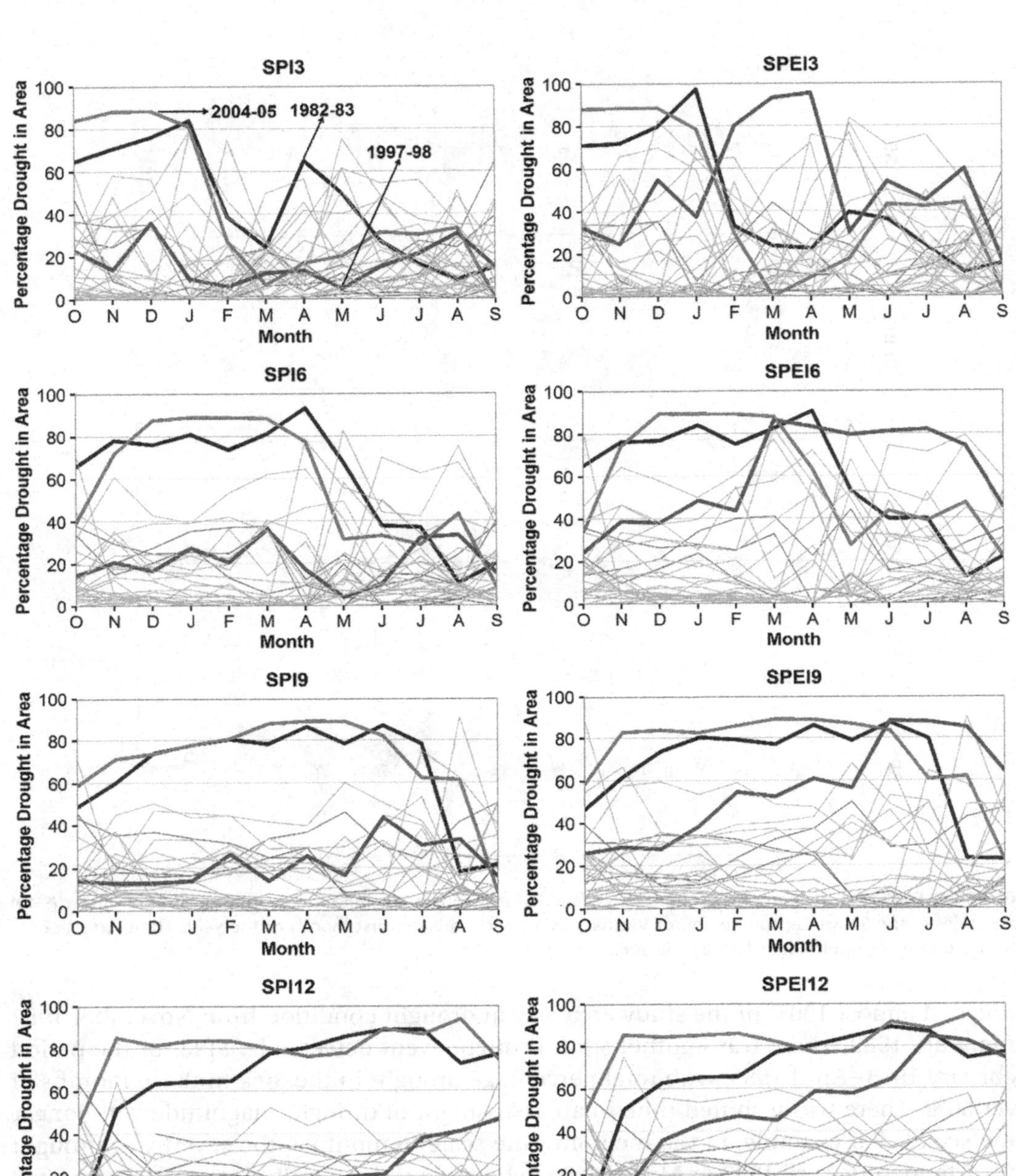

FIGURE 7.8 Comparison between percentage drought in area (PDA) based on the Standardized Precipitation Index (SPI) and that based on the Standardized Precipitation Evapotranspiration Index (SPEI) in the South Central Region of Vietnam during the period 1977–2014.

The progress of drought during 1997–98 was significantly different from the two other events. According to spatial assessment from SPEI3, drought started to affect a widespread area during the last months of 1997 to April 1998 and reduced afterward. A large PDA could be preserved until July with respect to medium time scale and long time scale.

To exploit the reason for making the different estimations between the SPI and SPEI for drought conditions in the year 1997–98, a further step was made by analyzing characteristics of rainfall and potential evapotranspiration during 1977–2014. Fig. 7.9 represents rainfall and temperature accumulations. Cumulative rainfall in the years 1982–83 and 2004–05 appeared in the lower bound as the first and second lowest cumulative rainfall, respectively. The total amount of rainfall received in those years was only equal to 64%–65% of that in normal years. Although total accumulated rainfall in the year 1997–98 was less than normal years (about 73%), it only ranked sixth among recorded data in terms of extreme low rainfall. This explains why using the precipitation-based SPI cannot highlight the year 1997–98 as an extreme case. From another perspective, potential evapotranspiration in 1997–98 showed an unusually high value. The estimated potential evapotranspiration in this period was 1934 mm, which was much higher than in normal conditions of 1734 mm. The cumulative potential evapotranspiration of 1982–83 and 2004–05, on the other hand, had a similar trend to that of normal evapotranspiration.

Cumulative climatic water balance anomalies, which were calculated by the difference between cumulative rainfall and that of potential evapotranspiration anomalies during the 38 years of the study area, are presented in Fig. 7.10. There was one strikingly different picture with rainfall accumulation. This was the year 1997–98 instead of the year 1982–83, taking the first lowest place, pushing the years 1982–83 and 2004–05 to second and third rank, respectively. The SPEI with its calculation based on accumulated

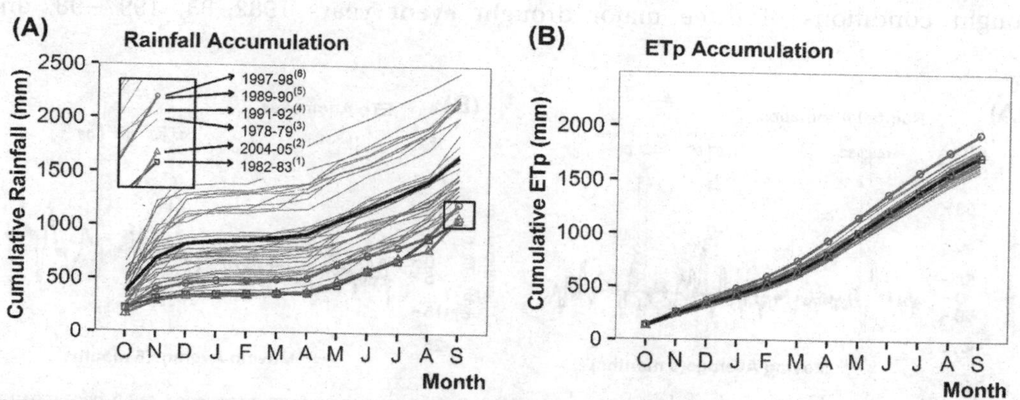

FIGURE 7.9 (A) Rainfall accumulation and (B) potential evapotranspiration accumulation in the South Central Region of Vietnam during the period 1977–2014. Superimposed with the *solid black line* represents the cumulative rainfall and potential evapotranspiration in normal years. The *gray lines* represent other periods. *ETp*, Potential evapotranspiration.

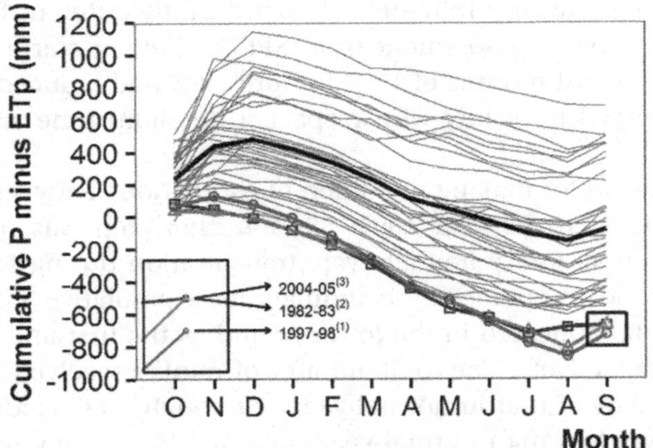

FIGURE 7.10 Cumulative difference between rainfall and potential evapotranspiration in the South Central Region of Vietnam during the period 1977–2014. Superimposed with the *solid black line* represents the cumulative of this difference in normal years. The *gray lines* represent other periods. *ETp*, Potential evapotranspiration.

climatic water balance therefore highlighted 1997–98 as an extreme occurrence of drought.

Another advantage of the SPEI is also highlighted through analyzing the trend of rainfall and potential evapotranspiration anomalies during the period 1977–2014 (Fig. 7.11). The linear regression line indicated a considerable upward trend for monthly potential evapotranspiration anomalies during the given time. It was inferred that temperature also has the same trend as potential evapotranspiration since this study used temperature solely to calculate potential evapotranspiration.

In short, through a comparison between the SPI and SPEI by assessing spatial drought conditions of three major drought event years 1982–83, 1997–98, and

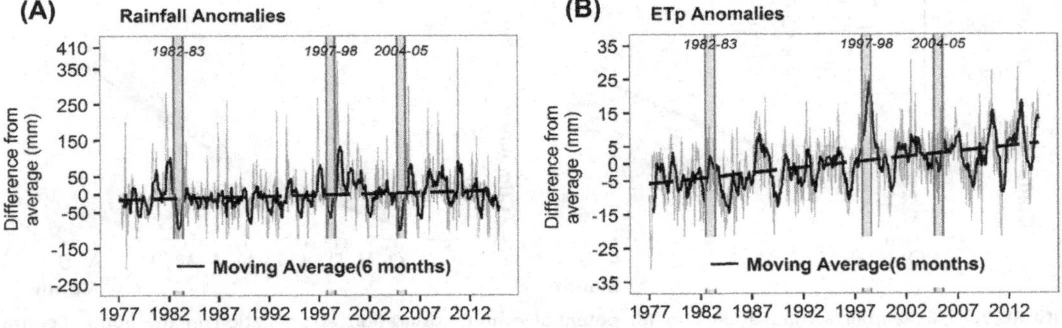

FIGURE 7.11 (A) Rainfall anomalies and (B) potential evapotranspiration anomalies trends during 1977–2014. *Bold black lines* are the moving averages of 6 months and *dashed linear lines* are produced by linear regression from time series. *ETp*, Potential evapotranspiration.

2004–05, it was concluded that both drought indices captured drought in the years 1982–83 and 2004–05, but only the SPEI reflected the severity of drought conditions in the year 1997–98. The reason is that the mechanism of drought in this year was a combination of a deficiency of rainfall and abnormally high temperature, causing the rainfall-based SPI not to highlight it as an extreme event. On the other hand, the SPEI considers both of the foregoing main factors in its calculation, leading to an acceptable estimation of the process of drought. Therefore the next section uses the SPEI for further discussion.

4.4 Characteristics of Moderate, Severe, and Extreme Drought

It is noticed that with the same PDA at a specific time, if a more severe drought condition dominates at a high portion, the consequences will be likely higher than that mainly dominated by less severe drought conditions. For this reason, a further analysis to investigate the characteristics of different types of drought in time series is essential. Fig. 7.12 shows time series of moderate, severe, and extreme drought conditions at four SPEI time scales during 38 years. This section uses water year to represent time series so there were some months of 1976–77 and 2014–15 that did not have values and were indicated in a gray color. The color-coded part presents the portion of total drought cells during a month. It ranges from 0% to 75%, which means that the variability of the PDA of the given study was remarkable. In spite of three historical drought years, there were many more drought years detected by color coding moderate droughts. It seems that on longer time scales, moderate drought affected the study area with a frequency of 1–2 years, while the recurrence time of severe and extreme drought was not clear. A high percentage of drought in an area was often observed during the winter–spring season and summer–autumn season at short time scales, while that of long time scales was in the summer–autumn season. Moreover, the PDA is likely irregular with many spikes. There may be no drought in previous months but in current months, the drought could be suddenly widespread in a large area.

Fig. 7.13 reveals a typical spatial drought event that happened during the year 2004–05 in the South Central Region of Vietnam by an analysis from SPEI6. The drought started to become severe from November 2004 in the whole of Binhthuan and Khanhhoa provinces and most of Phuyen province. Interestingly, the extreme drought area seems to appear first in the mountainous region at the western part of Khanhhoa province. Over the next 3 months, extreme drought swept across all four provinces except for the northern part of Binhdinh province. The drought cells were significantly high, with a range from 89% to 89.5% of total area. Although the drought cells remained high in March 2005 (88%) and April 2005 (63.3%), the magnitude of drought dropped from extreme to moderate and severe conditions. The drought situation continued to reduce over the following months excluding the coastal area of Ninhthuan province. This area was struggling with a persistent drought from severe to extreme conditions until the end of August 2005.

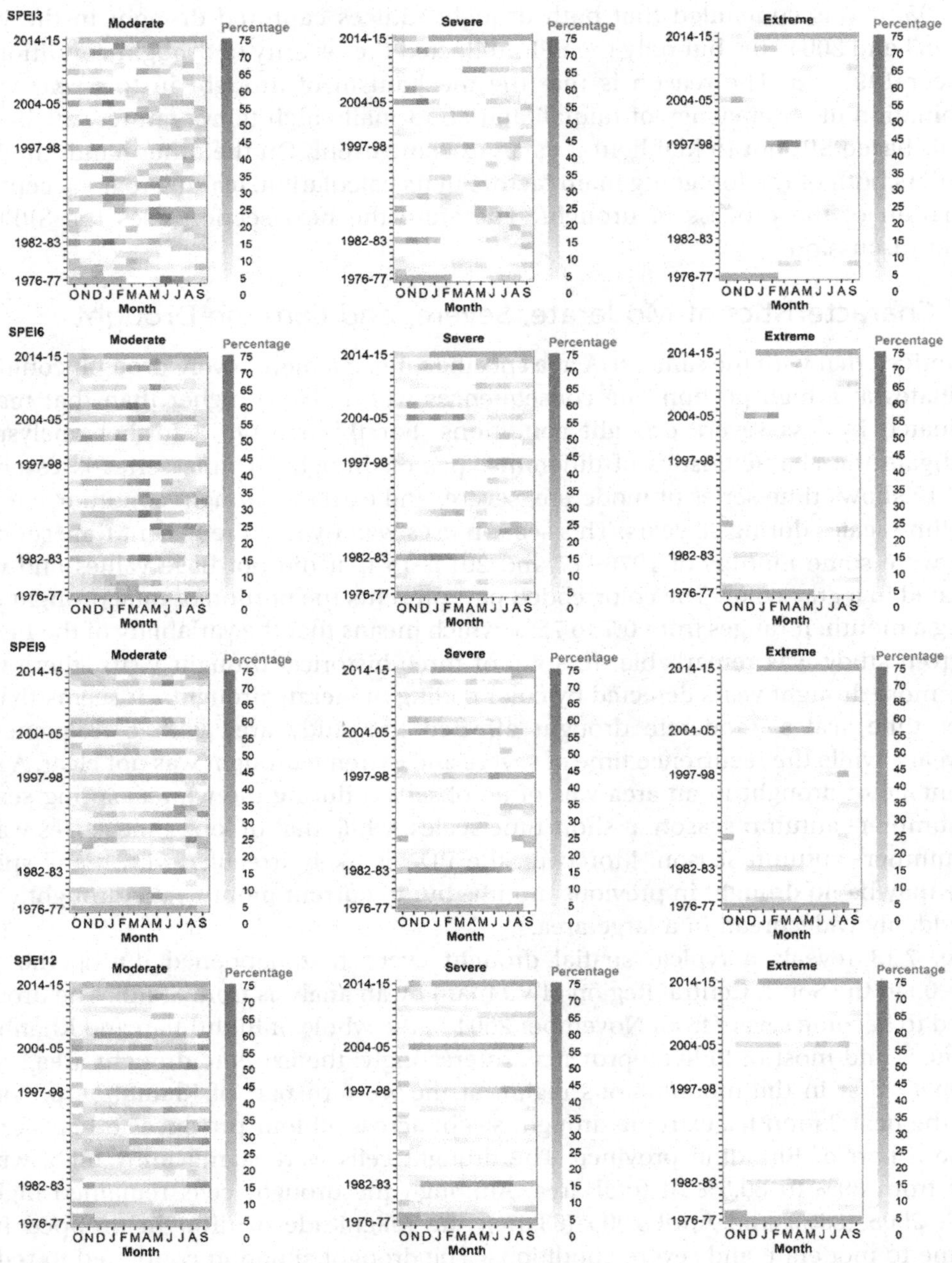

FIGURE 7.12 Time series of area in drought based on different time scales of the Standardized Precipitation Evapotranspiration Index (SPEI) in the South Central Region of Vietnam. The *gray color* indicates no value at that time.

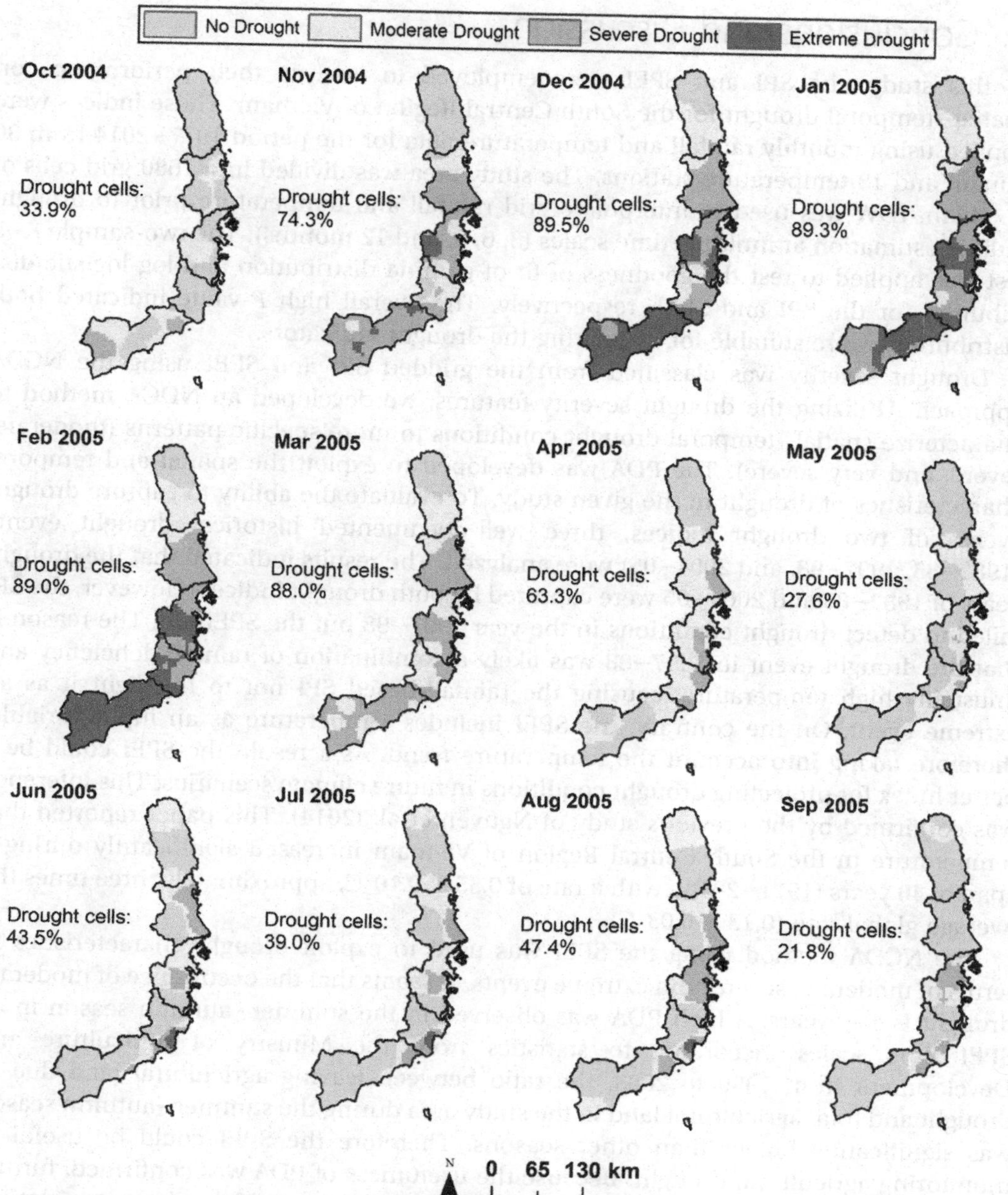

FIGURE 7.13 Spatial drought from October 2004 to September 2005 based on SPEI6 in the South Central Region of Vietnam.

5. Conclusions and Discussion

In this study, the SPI and SPEI were employed to analyze their performance on spatial–temporal drought for the South Central Region of Vietnam. Those indices were applied using monthly rainfall and temperature data for the period 1977–2014 from 30 rainfall and 13 temperature stations. The study area was divided into 1680 grid cells of 4×4 km. IDW was used to interpolate grid rainfall and temperature prior to drought indices estimation at multiple time scales (3, 6, 9, and 12 months). The two-sample K–S test was applied to test the goodness of fit of gamma distribution and log-logistic distribution for the SPI and SPEI, respectively. The overall high *P*-value indicated both distributions were suitable for estimating the drought indicators.

Drought severity was classified from the gridded SPI and SPEI using the NCDA approach. Utilizing the drought severity features, we developed an NDCA method to characterize spatial–temporal drought conditions to more specific patterns (moderate, severe, and very severe). The PDA was developed to exploit the spatial and temporal characteristics of drought in the given study. To evaluate the ability to capture drought events of two drought indices, three well-documented historical drought events (1982–83, 1997–98, and 2004–05) were analyzed. The results indicated that the drought years of 1982–83 and 2004–05 were captured by both drought indices. However, the SPI failed to detect drought conditions in the year 1997–98 but the SPEI did. The reason is that the drought event in 1997–98 was likely a combination of rainfall deficiency and unusually high temperature, causing the rainfall-based SPI not to highlight it as an extreme event. On the contrary, the SPEI includes temperature as an input variable, therefore taking into account the temperature trend. As a result, the SPEI could be a better index for projecting drought conditions in future climate scenarios. This inference was confirmed by the previous study of Nguyen et al. (2014). This paper reported that temperature in the South Central Region of Vietnam increased significantly during a span of 40 years (1971–2010), with a rate of $0.35 \pm 0.10°C$, approximately three times the average global rate ($0.13 \pm 0.03°C$).

The NCDA method using the SPEI was used to exploit drought characteristics in terms of moderate, severe, and extreme events. It seems that the occurrence of moderate drought is 1–2 years. A high PDA was observed in the summer–autumn season in all SPEI time scales. According to statistics from the Ministry of Agriculture and Development from 1980 to 2003, the ratio between leaving agricultural land due to drought and total agricultural land in the study area during the summer–autumn season was significantly higher than other seasons. Therefore the SPEI could be useful in monitoring agricultural drought. Because the usefulness of PDA was confirmed, further study could integrate moderate/severe/extreme drought types with land use, population, socioeconomic impact, and so on, to form a vulnerability and risk map. On the other hand, the PDA seems to be highly irregular with many spikes, causing difficulty in forecasting this measure.

Spatial drought propagation for the year 2004–05 was considered with SPEI6. The finding of this propagation is consistent with the work from Pham and Le (2009), which reported that Binhthuan province was one of the most vulnerable to drought during the last months of 2004. Other research from Nguyen and Shaw (2011), on the other hand, reported that the 2004–05 drought primarily affected Ninhthuan province first, then expanded to the whole region by the following year. The inconsistent conclusion about drought propagation in the study area therefore needs further investigation for the future.

This chapter mainly developed the NCDA method by focusing on drought magnitude but less on drought duration. There were a number of droughts that caused dire consequences such as those in 1993 and 2002, but were not clearly observed using either the SPI or SPEI. During these 2 years, the negative drought indices values were observed over a continuously long time (Figs. 7.6 and 7.7); however, only a small number were below the drought threshold of −1 to be defined as a drought state. Although having several limitations, the SPEI and NCDA method could be capable of constructing spatial–temporal droughts for the South Central Region of Vietnam, compared with a combination of the SPI and NCDA. This encourages consideration from multiple perspectives, for example, developing the NCDA method by combining both drought magnitude and duration. It is hoped that this broader view will improve our understanding of the nature of spatial–temporal drought.

Appendix 1: Characteristics Description of 30 Rainfall Measurement Stations (1977–2014)

Station Code	Station Name	Province	Latitude	Longitude	Elevation (m)	Annual Mean (mm)	Annual Maximum (mm)	Annual Minimum (mm)	SD (mm)
30701004	Binhtuong	Binhdinh	13°56′N	108°52′E	81	1872.0	3020.2	968.1	498.1
30701005	Bongson	Binhdinh	14°26′N	109°09′E	22	2281.3	3655.5	1213.2	573.1
30701008	Hoainhon	Binhdinh	14°31′N	109°02′E	9	2102.4	3481.4	1013.9	539.6
30701009	Phucat	Binhdinh	14°00′N	109°04′E	20	1897.9	3081.8	937.0	548.5
30701010	Phumy	Binhdinh	14°10′N	109°03′E	17	2037.7	3238.7	1131.9	546.9
30701011	Quynhon	Binhdinh	13°46′N	109°13′E	23	1893.4	3026.5	1130.7	489.2
30701014	Vancanh	Binhdinh	13°37′N	109°00′E	47	2155.6	3436.0	879.8	605.6
30702002	Cumong	Phuyen	13°40′N	109°11′E	14	2274.2	3490.7	1018.6	626.1
30702003	Cungson	Phuyen	13°02′N	108°59′E	42	1711.2	2901.7	934.2	500.5
30702011	Songcau	Phuyen	13°27′N	109°16′E	12	1803.0	2958.1	902.2	513.0
30702015	Tuyhoa	Phuyen	13°05′N	109°17′E	16	2062.6	3360.0	1030.7	559.0
30703001	Camranh	Khanhhoa	11°57′N	109°10′E	48	1250.4	2358.0	671.6	439.5
30703005	Dongtrang	Khanhhoa	12°17′N	109°02′E	19	1567.8	2648.8	784.8	476.1
30703007	Khanhson	Khanhhoa	11°59′N	108°56′E	408	1741.3	4175.1	766.1	714.6
30703008	Khanhvinh	Khanhhoa	12°17′N	108°54′E	31	1629.0	2913.3	679.9	600.2

30703009	Nhatrang	Khanhhoa	12°13′N	109°12′E	12	1435.0	2622.8	802.7	481.4
30703010	Ninhhoa	Khanhhoa	12°30′N	109°08′E	20	1486.2	2612.6	541.7	510.7
30704001	Bathap	Ninhthuan	11°42′N	109°03′E	19	814.0	1543.7	343.6	315.7
30704002	Cana	Ninhthuan	11°21′N	108°52′E	319	862.6	1999.5	401.9	356.3
30704006	Nhaho	Ninhthuan	11°40′N	108°54′E	36	853.5	1534.8	483.1	263.8
30704009	Phanrang	Ninhthuan	11°35′N	108°59′E	10	805.7	1633.9	449.1	296.5
30704011	Tanmy	Ninhthuan	11°43′N	108°48′E	44	1161.4	2100.1	665.5	347.2
30705003	Bautrang	Binhthuan	11°04′N	108°25′E	50	755.6	1293.6	346.7	197.7
30705008	Hamtan	Binhthuan	10°41′N	107°46′E	5	1622.9	2181.8	989.3	254.8
30705010	Langau	Binhthuan	11°11′N	107°47′E	139	2250.9	2888.1	1371.8	362.9
30705011	Lienhuong	Binhthuan	11°14′N	108°43′E	17	704.9	1266.0	245.4	242.0
30705016	Phanthiet	Binhthuan	10°56′N	108°06′E	7	1149.4	1768.1	784.9	208.6
30705017	Phuquy	Binhthuan	10°31′N	108°56′E	9	1296.5	2103.4	783.9	341.4
30705018	Songluy	Binhthuan	11°11′N	108°20′E	106	1097.3	1604.2	558.5	214.1
30705020	SuoiKiet	Binhthuan	11°03′N	107°42′E	113	1982.6	2533.1	1306.1	258.4

SD, standard deviation.

Appendix 2: Characteristics Description of 13 Temperature Measurement Stations (1977–2014)

Station Code	Station Name	Province	Latitude	Longitude	Elevation (m)	Annual Mean (°C)	Annual Maximum (°C)	Annual Minimum (°C)	SD (°C)
30701008	Hoainhon	Binhdinh	14°31′N	109°02′E	9	26.1	27.0	25.4	0.3
30701011	Quynhon	Binhdinh	13°46′N	109°13′E	23	27.1	28.0	26.4	0.3
30702003	Cungson	Phuyen	13°02′N	108°59′E	42	26.0	27.2	25.4	0.3
30702015	Tuyhoa	Phuyen	13°05′N	109°17′E	16	26.7	27.6	26.0	0.4
30703001	Camranh	Khanhhoa	11°57′N	109°10′E	48	27.1	27.9	26.4	0.4
30703005	Dongtrang	Khanhhoa	12°17′N	109°02′E	19	26.5	27.8	25.4	0.7
30703009	Nhatrang	Khanhhoa	12°13′N	109°12′E	12	26.8	27.5	26.2	0.3
30703010	Ninhhoa	Khanhhoa	12°30′N	109°08′E	20	27.6	28.5	26.6	0.4
30704006	Nhaho	Ninhthuan	11°40′N	108°54′E	36	27.3	28.1	26.7	0.3
30704009	Phanrang	Ninhthuan	11°35′N	108°59′E	10	27.0	27.6	26.4	0.3
30705008	Hamtan	Binhthuan	10°41′N	107°46′E	5	26.5	27.3	26.0	0.3
30705016	Phanthiet	Binhthuan	10°56′N	108°06′E	7	26.9	27.8	26.5	0.3
30705017	Phuquy	Binhthuan	10°31′N	108°56′E	9	27.2	28.3	26.7	0.3

SD, standard deviation.

Acknowledgments

This study is partially funded by project no. 101077 and Advanced Class in Translating Science to Application held at UNESCO-IHE, Delft, the Netherlands. The authors also thank reviewers for their valuable comments to improve the manuscript.

References

Bayissa, Y.A., et al., 2015. Spatio-temporal assessment of meteorological drought under the influence of varying record length: the case of upper Blue Nile Basin, Ethiopia. Hydrological Sciences Journal 60 (11), 1927–1942.

Blumenstock Jr., G., 1942. Drought in the United States Analyzed by Means of the Theory of Probability. US Department of Agriculture.

Corzo Perez, G.A., Van Huijgevoort, M., Voß, F., Van Lanen, H., 2011. On the spatio-temporal analysis of hydrological droughts from global hydrological models. Hydrology and Earth System Sciences 15, 2963–2978.

Dahal, P., et al., 2016. Drought risk assessment in central Nepal: temporal and spatial analysis. Natural Hazards 80 (3), 1913–1932.

Dao, X.H., 2002. Drought and its Mitigation Measures (In Vietnamese). Agricultural Publishing House, Hanoi, Vietnam.

Darling, D.A., 1957. The Kolmogorov-smirnov, cramer-von mises tests. The Annals of Mathematical Statistics 28 (4), 823–838.

Deo, R.C., Şahin, M., 2015. Application of the artificial neural network model for prediction of monthly standardized precipitation and evapotranspiration index using hydrometeorological parameters and climate indices in eastern Australia. Atmospheric Research 161, 65–81.

General Statistic Office of Vietnam, 2014. Population and Employment. Summary Report. Hanoi. URL: https://www.gso.gov.vn. access date: 06 Jan 2018.

Gobin, A., et al., 2012. Impact of Global Climate Change and Desertification on the Environment and Society in the Southern Centre of Vietnam (Case Study in the Binh Thuan Province).

IOM, 2014. Integrating Migration into Development: Diaspora as a Development Enabler. Summary Report, 2–3 October 2014. Italian Ministry of Foreign Affairs and International Cooperation, Rome.

Kelley, C.P., Mohtadi, S., Cane, M.A., Seager, R., Kushnir, Y., 2015. Climate change in the Fertile Crescent and implications of the recent Syrian drought. Proceedings of the National Academy of Sciences 112 (11), 3241–3246.

Le, M.H., Perez, G.C., Medina, V., Solomatine, D., 2017. Studying the impact of infilling techniques on drought estimation—a case study in the south central region of vietnam. In: Seventh International Conference on Information Science and Technology, Danang, Vietnam.

Lorenzo-Lacruz, J., et al., 2010. The impact of droughts and water management on various hydrological systems in the headwaters of the Tagus River (central Spain). Journal of Hydrology 386 (1), 13–26.

MARD, 2016. Drought Mitigation on Vietnam Southern Centre, Central Highland and Eastern South Regions by El Nino Phenomenon Report (In Vietnamese). Ministry of Agricultural and Rural Developement (MARD), Hanoi, Vietnam.

McEvoy, D.J., Huntington, J.L., Abatzoglou, J.T., Edwards, L.M., 2012. An evaluation of multiscalar drought indices in Nevada and eastern California. Earth Interactions 16 (18), 1–18.

McKee, T.B., Doesken, N.J., Kleist, J., 1993. The relationship of drought frequency and duration to time scales. In: Proceedings of the 8th Conference on Applied Climatology. American Meteorological Society Boston, MA, pp. 179–183.

Mishra, A., Desai, V., 2005. Spatial and temporal drought analysis in the Kansabati river basin, India. International Journal of River Basin Management 3 (1), 31–41.

Mishra, A., Singh, V.P., 2009. Analysis of drought severity-area-frequency curves using a general circulation model and scenario uncertainty. Journal of Geophysical Research: Atmospheres 114 (D6).

Mishra, A.K., Singh, V.P., 2010. A review of drought concepts. Journal of Hydrology 391 (1), 202–216.

Munger, T.T., 1916. Graphic method of representing and comparing drought INTENSITIES. 1. Monthly Weather Review 44 (11), 642–643.

Nguyen, D.Q., Renwick, J., McGregor, J., 2014. Variations of surface temperature and rainfall in Vietnam from 1971 to 2010. International Journal of Climatology 34 (1), 249–264.

Nguyen, H., Shaw, R., 2011. Chapter 8: Drought Risk Management in Vietnam, Droughts in Asian Monsoon Region. Emerald Group Publishing Limited, pp. 141–161.

Nguyen, L.B., Li, Q.F., Ngoc, T.A., Hiramatsu, K., 2015. Drought assessment in Cai River basin, Vietnam: a comparison with regard to SPI, SPEI, SSI, and SIDI. Journal of the Faculty of Agriculture, Kyushu University 60 (2), 417–425.

Nguyen, Q.K., 2005. Drought Investigation and Risk Reduction on Vietnam South Central Region and Central Highlands (In Vietnamese). Vietnam Ministry of Science and Technology, Ho Chi Minh City.

Palmer, W.C., 1965. Meteorological Drought, 30. US Department of Commerce, Weather Bureau Washington, DC.

Pham, Q.H., Le, D.T., 2009. Water resources in dry season of Southern Central Vietnam with drought and sandy desert (from Phuyen to Binhthuan province) (in Vietnamese). Journal of Water Resources and Environmental Engineering 25, 3–8.

Potop, V., Možný, M., Soukup, J., 2012. Drought evolution at various time scales in the lowland regions and their impact on vegetable crops in the Czech Republic. Agricultural and Forest Meteorology 156, 121–133.

Quiring, S.M., 2009. Developing objective operational definitions for monitoring drought. Journal of Applied Meteorology and Climatology 48 (6), 1217–1229.

Rhee, J., Carbone, G.J., 2011. Estimating drought conditions for regions with limited precipitation data. Journal of Applied Meteorology and Climatology 50 (3), 548–559.

Sager, T.W., 2010. Kolmogorov-Smirnov test. In: Encyclopedia of Research Design. SAGE Publications, Thousand Oaks, CA, USA.

Santos, J.F., Pulido Calvo, I., Portela, M.M., 2010. Spatial and temporal variability of droughts in Portugal. Water Resources Research 46 (3).

Shaw, R., Prabhakar, S.V.R.K., Nguyen, N., Provash, M., 2007. Drought Management Considerations for Climate Change Adaption: Focus on the Mekong Region. Oxfam Vietnam and International Environment and Disaster Management Laboratory of Kyoto University.

Sienz, F., Bothe, O., Fraedrich, K., 2012. Monitoring and quantifying future climate projections of dryness and wetness extremes: SPI bias. Hydrology and Earth System Sciences 16 (7), 2143.

Stagge, J.H., Tallaksen, L.M., Gudmundsson, L., Van Loon, A.F., Stahl, K., 2015. Candidate distributions for climatological drought indices (SPI and SPEI). International Journal of Climatology 35 (13), 4027–4040.

The Socialist Republic of Vietnam, 2004. National Report on Disaster reduction in Vietnam, for World Conference on Disaster Reduction Kobe-Hyogo. Japan. Hanoi, URL: https://www.unisdr.org/2005/mdgs-drr/national-reports/Vietnam-report.pdf, access date: 12 December 2017.

Törnros, T., Menzel, L., 2014. Addressing drought conditions under current and future climates in the Jordan River region. Hydrology and Earth System Sciences 18 (1), 305.

Tran, T.L.H., 2016. Water resource for economic development in Vietnam and implications for developing countries. Global Journal of Management and Business Research 15 (11).

UNESCAP, 2015. Overview of Natural Disasters and Their Impacts in Asia and the Pacific 1970-2014, ICT and Disaster Risk Reduction Division.

Vicente-Serrano, S.M., 2006. Differences in spatial patterns of drought on different time scales: an analysis of the Iberian Peninsula. Water Resources Management 20 (1), 37–60.

Vicente-Serrano, S.M., Beguería, S., López-Moreno, J.I., 2010. A multiscalar drought index sensitive to global warming: the standardized precipitation evapotranspiration index. Journal of Climate 23 (7), 1696–1718.

Vicente-Serrano, S.M., et al., 2012. Performance of drought indices for ecological, agricultural, and hydrological applications. Earth Interactions 16 (10), 1–27.

Vicente-Serrano, S.M., López-Moreno, J.I., 2005. Hydrological response to different time scales of climatological drought: an evaluation of the Standardized Precipitation Index in a mountainous Mediterranean basin. Hydrology and Earth System Sciences Discussions 9 (5), 523–533.

Vicente-Serrano, S.M., National Center for Atmospheric Research Staff, 2015. The Climate Data Guide: Standardized Precipitation Evapotranspiration Index (SPEI).

Vu-Thanh, H., Ngo-Duc, T., Phan-Van, T., 2014. Evolution of meteorological drought characteristics in Vietnam during the 1961–2007 period. Theoretical and Applied Climatology 118 (3), 367–375.

Vu, M.T., Raghavan, S.V., Pham, D.M., Liong, S.-Y., 2015. Investigating drought over the Central Highland, Vietnam, using regional climate models. Journal of Hydrology 526, 265–273.

Wells, N., Goddard, S., Hayes, M.J., 2004. A self-calibrating Palmer drought severity index. Journal of Climate 17 (12), 2335–2351.

Wilhite, D.A., 2000. Drought as a Natural Hazard: Concepts and Definitions.

Wilhite, D.A., Glantz, M.H., 1985. Understanding: the drought phenomenon: the role of definitions. Water International 10 (3), 111–120.

WMO, 2009. Experts Agree on a Universal Drought Index to Cope with Climate Risks. WMO Press Release No. 872 http://www.wmo.int/pages/prog/wcp/agm/meetingswies09/documents/872_en.pdf.

WMO, 2012. Standardized Precipitation Index User Guide.

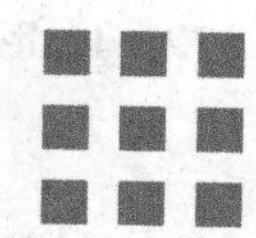

Index

'*Note*: Page numbers followed by "f" indicate figures, "t" indicate tables.'